T0290359

The Physics of Sound and Music, Volume 2

A complete course text (Lab manual)

Online at: https://doi.org/10.1088/978-0-7503-6350-1

The Physics of Sound and Music, Volume 2

A complete course text (Lab manual)

Samya Bano Zain
Department of Physics, Susquehanna University, Selinsgrove, PA, USA

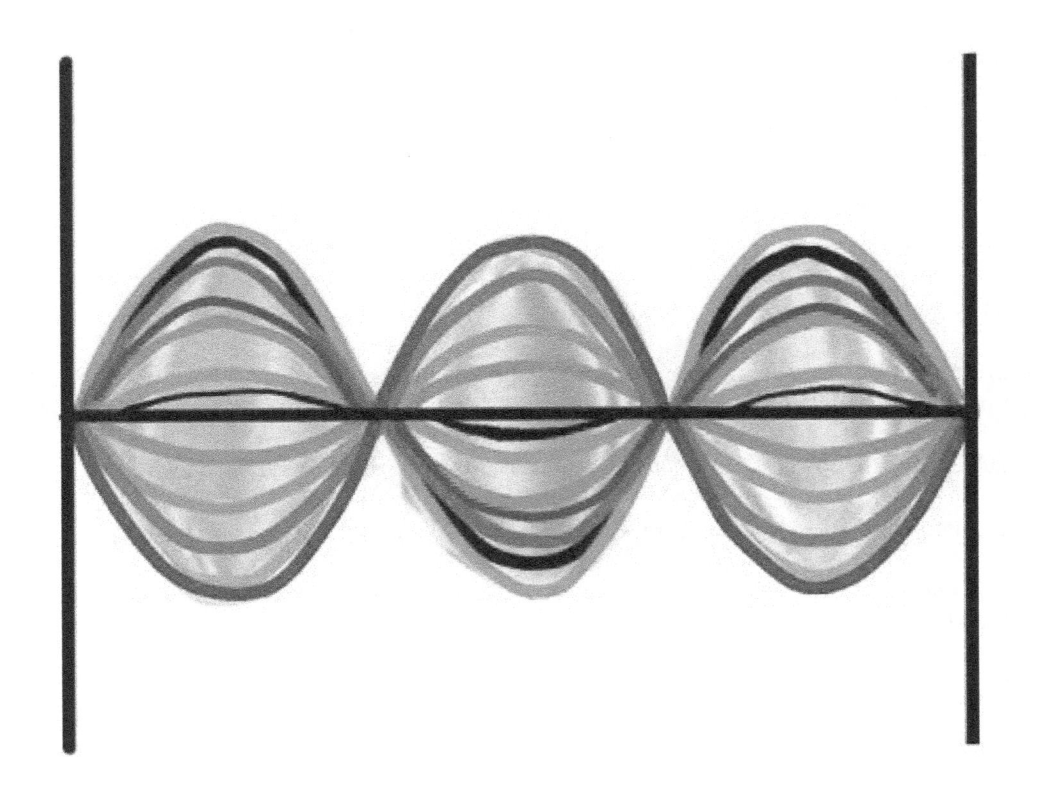

IOP Publishing, Bristol, UK

ISBN 978-0-7503-6350-1 (ebook)
ISBN 978-0-7503-6348-8 (print)
ISBN 978-0-7503-6351-8 (myPrint)
ISBN 978-0-7503-6349-5 (mobi)

DOI 10.1088/978-0-7503-6350-1

Version: 20240401

IOP ebooks

British Library Cataloguing-in-Publication Data: A catalogue record for this book is available from the British Library.

Published by IOP Publishing, wholly owned by The Institute of Physics, London

IOP Publishing, No.2 The Distillery, Glassfields, Avon Street, Bristol, BS2 0GR, UK

US Office: IOP Publishing, Inc., 190 North Independence Mall West, Suite 601, Philadelphia, PA 19106, USA

To my husband,

because everyone should have somebody to stand up for them.

Contents

Preface

The Physics of Sound and Music is, as the name implies, intended to be an introductory course in learning about the science behind sound and music. This book is the outcome of a physics course at Susquehanna University mainly for non-majors to fulfill their scientific explanations portion of the central curriculum. This book has sufficient material for a one semester four-credit hour-long course.

I recognize that mathematics is fundamental and indispensable to the study of physics. A little bit of familiarity with algebra is required to succeed in the course. However, I have avoided derivations of complex equations in the text. I have tried to establish, wherever possible, context and background. I feel that this manner of learning makes it easier for the reader to understand and grasp the validity of the particular physics concept/topic. For details of mathematical proofs of concepts please check other sources.

Main ideas for understanding

As with my other books, when writing this book I pictured myself conversing with a student and explaining the fundamental physics behind sound and its propagation. For this reason I have tried to keep it as conversational and down-to-earth as possible by (in part) avoiding too much unnecessary technical jargon. Readers should be able to pick it up and learn on their own. Worked examples are included as part of the text in nearly every section. Additional concept questions and exercises are included for further practice and understanding.

Mainly for instructors

In this text I have put physics concepts before the historic development of the field of acoustics. Many concepts are repeated accross multiple chapters. This repetition is intentional. I feel that reviews and re-reviews of important concepts never hurt. Some questions included at the end of sections may go just beyond the material in the chapter. Such content is intended to allow students to apply concepts they have learnt during the chapter and the book overall and hopefully increase their knowledge and undertanding.

I have incorporated a lab manual with 'activities-to-do' as a part of this textbook. I have tested most of these with students on our campus and I have found that they are useful in clarifying ideas that might be a little hard for the non-physics majors to comprehend and/or appreciate. A majority of activities included in this textbook are also intentionally kept simple and most can be done with materials that are easily acquired at most general stores and many do not require specialized physics or engineering equipment. The intention is to make these activities accessible to everyone around the world and make it easy for instructors to incorporate them in their own class and lab rooms as they see fit.

Mainly for students

Before we begin, I just have a few additional things to comment on. Remember, Physics must *not* be memorized but rather understood. Know that understanding does not come quickly or easily, it takes patience and resilience. Remember how you learnt to play an instrument? Did you become an expert overnight? Or did it take hours and hours of practice? Physics is just the same, you need to actively learn it, cramming overnight before the exam is not be the ticket to understanding and hence success.

Key to success

As always, my advice throughout the years has not changed.

1. Manage your time and get organized.
2. Commit yourself. Be a good listener and take good notes.
3. Find a study partner.
4. Motivate your partner and yourself.
5. Actively seek help.
6. Set a dedicated time to study physics every day. Thirty minutes a day, every day will get you more than pulling-an-all-nighter the night before a quiz.

And the most important of all, remember what Sam Ewing said: ‘Hard work spotlights the character of people; some turn up their sleeves, some turn up their noses and some don’t turn up at all.’ Hence, just show up to class, be a good listener, ask questions, and take good notes. Read the book, do the examples and exercises, share your knowledge, communicate and most important of all, work hard and enjoy the ride.

Samya Bano Zain
Professor, Susquehanna University
August 30, 2023

Acknowledgements

This work is based on over a decade of study, reflection and critique and hopefully it represents a marked improvement in my understanding and teaching about acoustics. For that I am grateful to the contributions of many, many people. They include, but are not limited to the numerous Susquehanna students I have had the pleasure to learn from, Dr Grosse, who first encouraged me to put words to paper, Dr Ken Brakke, Robert Everly, Dr Jeffrey Graham, for multiple informative and valuable physics and math conversations. I am also grateful to Dr Patrick Long, Department of Music, who graciously allowed me to attend his music theory class. A lot of technical musical improvements in this text are a direct result of his class and his wonderful teaching skills and I want to especially thank M L Klotz, Department of Psychology, for keeping me sane!

To my ever patient husband, Zain and my kids, Hareem, Sami and Eman, for being (nearly always!) willing models and getting excited about whatever crazy idea I had for a picture that particular day! and to Fahmida, the newest addition to our family. A special thank you to Sami's new phone and its picture quality and last but not least, Hareem and Eman's contributions are truly appreciated, thank you both for making artwork for me and letting me use your pictures in the book, it is deeply appreciated.

My heartfelt gratitude also extends to my family, my dad Dr Iqbal and my mom Safia, who live in my heart, my brother Dr Aamir Iqbal, and my sister-in-law Sofia Aamir. I also want to thank my relatives-in-law, Rauf bhai, Hifza baji and Shazia, and Shoaib bhai, who welcomed me to their family almost 30 years ago and have become more than my brothers and sisters over the years. You have always believed in me, prayed for me and encouraged me to be my best self. Thank you all for your unwavering support!

I am reminded of the quote by Sir Isaac Newton, '*If I have seen further than others, it is by standing upon the shoulders of giants*', I guess I have met many giants in my life and for that I am eternally grateful!

Author biography

Samya Bano Zain

Samya Bano Zain grew up in Lahore, Pakistan. In seventh grade she decided that she wanted to learn more physics, since physics explains how the Universe works, and additionally allowed her not to go into the medical field like the rest of her family. She completed her undergraduate work at the University of the Punjab, Lahore, Pakistan and was awarded the gold medal for the highest score in the annual final exam. Then with her husband and four-year-old son she went on to graduate work in physics at the university at Albany, State University of New York in 2001. She received her PhD in experimental particle physics in 2006 and continued as a post-doctoral fellow at the Albany High Energy Physics group. In the fall of 2008, she joined Susquehanna University as an assistant professor, where she was promoted to the rank of full professor last year in the Department of Physics.

When not working on physics she can often be found with her nose in a book. She makes time to read every day (even when she must hide from her family in order to do so). She paints when the mood strikes her and tries to create Urdu and Arabic calligraphy (not as well as she would like!), embroiders to relax and has recently taken up learning how to do mandala art. She loves visiting used bookstores and cannot pass up procuring more books and absolutely refuses to let any of her collection go!

Part I

Introduction

IOP Publishing

The Physics of Sound and Music, Volume 2
A complete course text (Lab manual)
Samya Bano Zain

Chapter 1

Introduction

1.1 Activity-to-do: review of mathematics

1.1.1 Part 1: Simple mathematics

1. $12 + 24 =$

2. $-16 + 8 =$

3. $-2 - 6 =$

4. $\frac{20}{-4} =$

5. $5 \times 2 \times (-2) =$

doi:10.1088/978-0-7503-6350-1ch1

1.1.2 Part 2: Least common multiple

When a number is multiplied by another number we get multiples of the number.
The common multiples are those that are found in two different numbers.
Least common multiple is the smallest of the common multiples.

Example 1.1. Multiples of 2: 2, 4, 6, 8, 10, 12, ...
Multiples of 3: 3, 6, 9, 12, ...
Common multiples: 6, 12, ...
Least common multiples of 2 and 3: LCM(2, 3) = 6

1. LCM(3, 6, 18) =

2. LCM(2,8,12, 20) =

3. LCM(4,15, 25) =

1.1.3 Part 3: Fractions

In a fraction the top number is called a numerator and the bottom number is called the denominator.

$$\text{Fraction} = \frac{\text{Numerator}}{\text{Denominator}} \tag{1.1}$$

To add or subtract fractions we make the denominators of both (or all) fractions the same by using the least common number method explained in Part 2. To multiply fractions we multiply numerators and multiply denominators of each number and then simplify the factions as needed. To divide factions, we first write the reciprocal of the second number and then use the rules for multiplication of fractions to calculate the answer.

1. $\frac{1}{3} + \frac{6}{3} =$

2. $\frac{9}{14} + \frac{4}{7} =$

3. $\frac{3}{8} \times 2\frac{3}{4} =$

4. $\frac{2}{8} \div \frac{3}{4} =$

1.1.4 Part 4: Exponents

Exponents are used as a short hand to express how many times a number needs to be multiplied in the expression. Exponents are also called *powers* or *indices*.

Examples: $2^5 = 2 \times 2 \times 2 \times 2 \times 2 = 32$, or $10^4 = 10 \times 10 \times 10 \times 10 = 10\ 000$

Negative exponents tell us how many times to divide the number:

Example: $5^{-3} = 1 \div 5^3 = \frac{1}{5 \times 5 \times 5} = \frac{1}{125} = 0.008$

1. $10^6 =$

2. $10^{-4} =$

3. $(3 \times 10^9) + (2 \times 10^4) =$

4. $(2.1 \times 10^4) - (3.1 \times 10^3) =$

5. $(5.5 \times 10^3) \times (2 \times 10^5) =$

6. $(9 \times 10^4) \div (3 \times 10^8) =$

1.1.5 Part 5: Algebra

Elementary algebra is essential for any study of mathematics, science, or engineering. Algebra is derived from Arabic '*al-jabar*' and is the study of mathematics using letters as stand-ins for numbers. Algebra defines the rules for utilizing these symbols in mathematical operations.

For example:

1. If $a = 2$, $b = 3$, $c = 4$, $d = 5$, then,
   (a) $2a + 4b + 3c =$

   (b) $-8a + 4(3b - c) =$

   (c) $\frac{a + 4b}{c - 3d} =$

2. Solve for the variable:

   (a) $-5x + 7(x - 2) = 0$

   (b) $-8y - 4(3y - 2) = 3$

   (c) $\frac{b - 2}{b - 3} = \frac{3}{4}$

1.1.6 Part 6: Problem solving

1. What is 30% of 20?

2. 16% of what number is 12?

3. 17 is what percent of 51?

4. 25 is what percent of 300?

5. Find 7/8 of 20

6. 80 people took a physics test. 70 of them passed the test. What percentage failed the test?

7. Mary managed to sell 3/5 of her cakes.
   (a) What percentage of cakes was sold?

   (b) What percentage of cakes was left?

8. Pam bought 50 kg of sugar. She used 10 kg of sugar. What percentage of sugar was left?

9. If you are running at 15 miles per hour, how far will you be after 2 hours?

1.2 Scientific method

Scientific method has been in effect since the 17th century. It is a general procedure for experimentation used to explore, observe and answer questions. The main purpose of scientific method is to discover cause and effect relationships. A generalized scientific method must consist of the following minimum number of steps, given in figure 1.1.

1. **Develop a question.** The first step is to develop a question about something you are interested in. It could be a direct observation or an indirect observation but it has to be posed as an inquiry: what, when, how, where etc.
2. **Hypothesis.** The next task is to develop a hypothesis, basically an educated guess. A hypothesis then allows you to make some predictions about your posed question.
3. **Background research.** Another way to ensure that your hypothesis is an educated guess is to conduct some background research about your question.

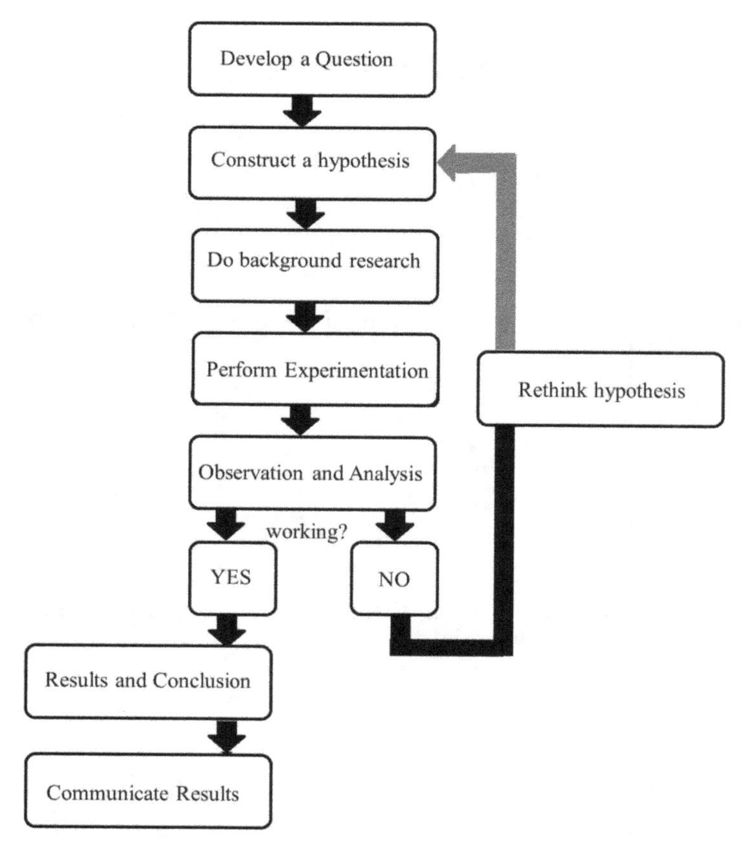

Figure 1.1. Scientific method in a box.

4. **Set up the experiment.** Next comes the task of experimentation. In order to test your hypothesis you must come up with an experiment that allows you to answer your question. Here you have to do the following:
   (a) Choose correct variables.
   (b) Choose appropriate apparatus.
   (c) Choose proper units.

   It is vital at this point to recognize that your experiment must be repeatable and be performed while varying as few variables as possible.
5. **Observation, recording data and analysis.** Your experiment must be followed by data collection and data analysis to test if the experiment supports your hypothesis or not. This must be done very precisely and accurately. Sometimes this step will lead to good results and conclusions that validate your hypothesis. Other times this may lead to the development of a completely new question and a new hypothesis. Both of which are desirable outcomes for a scientist.
6. **Results and conclusion.** Whether your hypothesis was supported by data or not.
7. **Communication.** Results are generally communicated to the scientific community by publishing it as a final report in a scientific journal or as a presentation at a scientific meeting.

1.2.1 Activity-to-do: scientific method

Choose to answer one of the following questions to do in-class.

1. How many of each color of M&Ms are in the bag?
2. Do some M&Ms weigh more than other M&Ms based on their color coating?
3. Which M&Ms colors dissolve the fastest?
4. Does each bag of M&Ms have the same number of each color?
5. What happens when you put hot versus cold water on M&Ms?

For this activity we will need the following materials.

1. M&Ms candy.
2. Plastic cups.
3. Stop watch.
4. Graph paper or Excel spreadsheet.

Example 1.2.

1. **Sample question.** How many of each color of M&Ms are in the bag?
2. **Sample hypothesis.** Practice your scientific thinking. Write down what you believe the answer is (take a guess).
3. **Sample background research.** Discuss within your team members how to approach/find the solution for this experiment. Questions to answer may include:
   (a) Guess how many actual M&Ms colors there are in the bag.
   (b) Will each bag have the same number of colors?
4. **Sample experimentation.**
   (a) Write the materials used for this experiment.
   (b) Sort M&Ms by different colors.
   (c) Count each color.
5. **Sample observations and analysis.**
   1. Develop a proper data table based on the experimental requirements. A data table is one type of graphic organizer used to represent qualitative and/or quantitative data. A proper table must include the following:
      (a) Title your table that directly relates to your experimental data.
      (b) Determine the number of columns and rows required.
      (c) Label your columns. The leftmost column is reserved for your independent variable.
      (d) Use proper units per column.
      (e) Record the data from your experiment in the newly created table (table 1.1).

Table 1.1. Number of colored M&Ms.

Color	Red	Green	Blue	Brown	Orange	Yellow	Total %	Class %
Number								

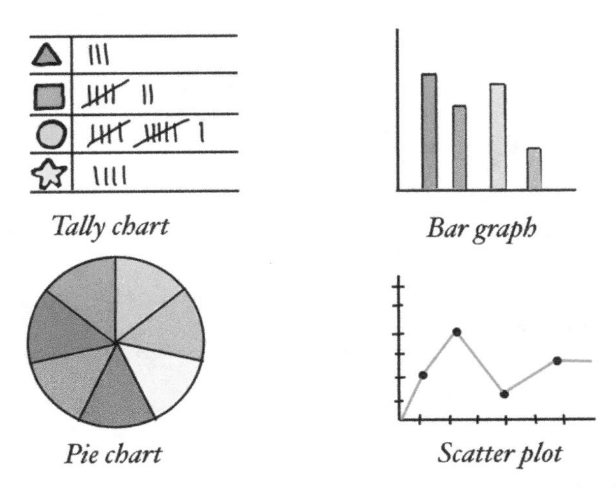

Figure 1.2. Sample graph papers.

2. Make different types of graphs commonly used for the overall proportions of M&M candies. The graph could be done on Excel or be made using pen and paper. The main types of graphs include (figure 1.2):
   (a) tally graph;
   (b) scatter plot;
   (c) bar graph;
   (d) pie chart.

6. **Sample Results and conclusions.**
   The conclusion represents the results you got and your interpretation of what they mean.
   (a) What did you observe?
   (b) Was your hypothesis correct?
   (c) What did you learn?
   (d) What could be some sources of errors?
   (e) If you were to do the experiment again, what changes would you make?

1.2.1.1 Activity-to-do: how many of each color of M&Ms?

1. **Develop a question.**

2. **Hypothesis.**

3. **Background research.**

4. **Experimentation.**

5. **Observation and analysis.**

6. **Results and conclusion.**

1.3 Units

Physics is an experimental science and experiments involve complex measurements. Surprisingly, complex measurements can be expressed using a few fundamental quantities, like time (t), length (L), and mass (m). *'Fundamental units'*, or SI units, abbreviated from the French, *Le Systeme International d'Unites* (translated as the International System of Units) uses seconds, meters, and kilograms in physical experiments.

Example 1.3.

1. There are 12 inches in 1 foot. How many feet in 48 inches?
Solution:

2. There are 1.6 km in a mile. How many miles in a 10 K race?
Solution:

3. How many seconds are there in one year?
Solution:

1.3.1 Derived units

The International System of Units (SI) specifies a set of seven base units. From these seven base units other SI units of measurement are derived. We call these the *'derived units'*. Each derived unit can be expressed as a product of one or more base units. For example, the SI derived unit of area is the square meter (m^2) and that of density is kilogram per meter cubed. (kg m^{-3}).

1.3.1.1 Speed

Speed is how fast something moves and is defined in physics as the rate of change of position of an object in a particular direction. Speed is measured as the ratio of distance to the time in which the distance was covered. Mathematically, speed is expressed as:

$$\text{Speed} = \frac{\Delta x}{\Delta t} = \frac{\text{m}}{\text{s}}, \tag{1.2}$$

where Δx is the change in distance and Δt is the change in time.

Example 1.4. Samantha runs 360 meters in 1 minute. What is her speed in m s^{-1}?
Solution:

1.3.1.2 Activity-to-do: constant walking speed

1. Measure and mark with a masking tape 10 meters in a hall.
2. Walk the distance while a friend keeps time.
3. What is your average speed?
4. Is there an easy way that you could change your speed? (Hint: what do the terms acceleration and deceleration mean in English?)

1.3.1.3 Activity-to-do: constant velocity car—a two people activity

Materials needed:

Motorized car (one with constant speed); scissors; streamer; paper; marking tape; meter stick.

To do:

1. Attach a long streamer (or paper) with tape to the back of the car.
2. Put the motorized car at a user-defined origin.
3. Sit very close to the car and let go of the car.
4. As the car moves put dots on the streamer every second, keeping the dots as consistent as possible.
5. Cut the streamer (or paper) at the dots.
6. Graph distance versus time and speed versus time (figure 1.3), what do you observe?

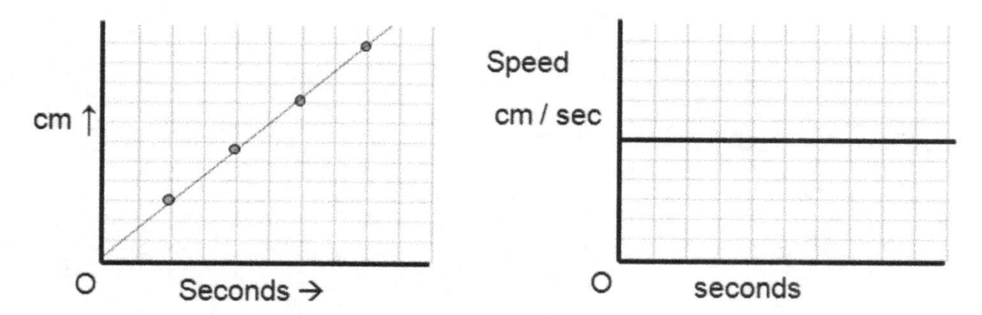

Figure 1.3. Graph for object with a constant speed.

1.3.1.4 Activity-to-do: constant velocity car—multiple people—class activity

Materials needed:

Motorized car (one with constant speed, one with accelerating motion); paper; marking tape; meter stick for marking; stopwatches.

To do:

1. Mark out 10 meters; from 0 to 10 m.
2. Put the motorized car at the 0 mark.
3. Have each person stand at the 50 cm mark with their stopwatches.
4. Have one person sit and let go of the car.
5. As the car moves by each person, stop the timer and keep track of time.
6. Repeat the experiment with both cars three times.
7. Put the values the class got on the board, in the order of increasing distances.
8. Graph distance versus time and speed versus time for the three trials (figure 1.4). Hint: Excel can be used for graphing.
9. What do you observe?

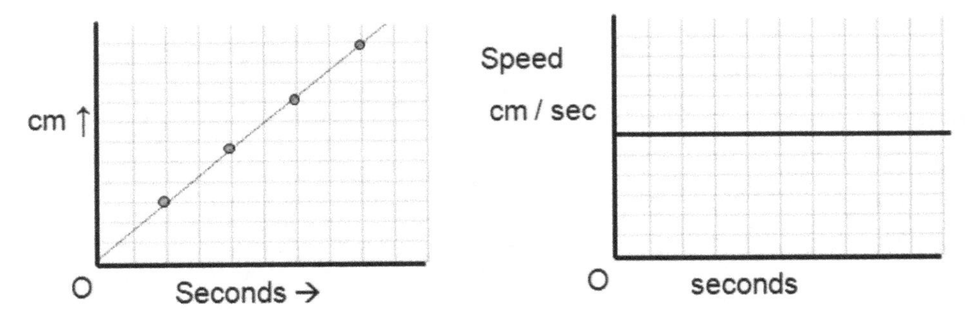

Figure 1.4. Graph for object with a constant speed.

1.4 A few important concepts

Before we begin, a formal discussion: let us first define a few important concepts we will use throughout the text.

1. **Coordinate system:** a coordinate system is an artificially imposed grid that allows us to solve problems by making quantitative measurements.
2. **Origin:** an origin is an agreed upon reference point in a particular coordinate system. All measurements are made with respect to the origin.
3. **Three dimensions of space:** we live in a three-dimensional (3D) space, it is obvious that we will require three coordinates to locate an object relative to a user-defined origin. In general we use Cartesian coordinates (x, y, z), but we also use spherical and cylindrical coordinates.

1.4.1 Activity-to-do: make a 3D coordinate system with paper

Materials needed:

Graph paper; pen; scissors; clear tape.

To do:

1. Put graph paper on the table.
2. Fold the paper over into a triangle and cut the extra bit left over.
3. Open the paper and put flat on the table.
4. Fold the paper into two equal rectangles, so you have four squares.
5. At the middle of the paper mark the origin (O).
6. Mark along two folds, one horizontal and one vertical, starting from origin to ends of the paper.
7. Use the scissors to cut along one line until the middle, do not cut beyond the middle of the paper.
8. Fold and tape along the cut.
9. You have just created a 3D coordinate system!

1.5 Review of vectors

1.5.1 Activity-to-do: understanding vectors—battleships game!

Materials needed:
Battleships game; graph paper; pen; calculator.

Experiment-to-do:

1. Play the game and see what you find.

1.5.2 Scalars

In physics, quantities that can be fully described by a numerical value are called scalar quantities. A scalar quantity has a magnitude (size) but no direction. For example the money in your pocket, your age, your mass, and the temperature of a room are scalar quantities.

1.5.3 Vectors

Quantities that need direction along with magnitude to completely describe them are called vector quantities. In other words, vector quantities are scalars with a direction. Familiar examples of a vector include velocity, acceleration and force.

A given vector $\vec{A}$ is specified by defining its magnitude and direction relative to a chosen reference frame.

$$\vec{A} = (a_x, a_y, a_z) = a_x\hat{\imath} + a_y\hat{\jmath} + a_z\hat{k} \tag{1.3}$$

If $\vec{v}$ is a velocity vector then $\vec{v}_x$ is the velocity of the particle along the x-direction; $\vec{v}_y$ is the velocity of the particle along the y-direction; $\vec{v}_z$ is the velocity of the particle along the z-direction.

1.5.4 Distance versus displacement

Distance: distance is a scalar quantity. It is a numerical description of how far apart two objects are. In other words the actual physical path you traversed from initial to final position is called distance traveled.

Displacement: displacement is the shortest distance from the initial to the final point. Displacement is a vector quantity and hence it has a magnitude as well as direction. The displacement vector is determined by the starting and ending points of the interval traversed and not by the physical path traveled between the two points.

1.5.5 Components of a vector

A vector is a quantity described by length and direction. It is also useful to describe a vector in terms of *its components*, especially in situations when dealing with vectors mathematically. A vector $\vec{a}$ in figure 1.5. is described by a magnitude or length $|\vec{a}|$ from a specific user-defined origin, and its direction is specified by an angle θ, measured with respect to (w.r.t.) a user-defined reference line, generally the positive x-axis. The x- and y-components of the vector $\vec{a}$ are mathematically written as,

$$\begin{cases} a_x = a\cos\theta \\ a_y = a\sin\theta \end{cases} \tag{1.4}$$

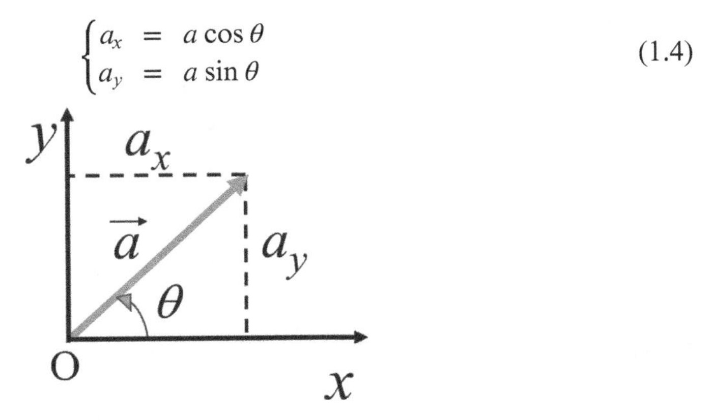

Figure 1.5. Components of a vector.

In this form the vector $\vec{a}$ is a vector sum of the two components,

$$\vec{a} = a_x\hat{i} + a_y\hat{j} \quad (1.5)$$

where $\hat{i}$ and $\hat{j}$ are unit vectors along the x and the y-axis.

1.5.6 Adding vectors

Pythagoras, a Greek mathematician (570–495 BCE) is credited for the discovery and proof of the Pythagorean theorem, which allows us to add vectors. The 'net' displacement $\vec{C}$ is equal to the first displacement $\vec{A}$ plus the second displacement $\vec{B}$. Mathematically written as,

$$\vec{C} = \vec{A} + \vec{B} \quad (1.6)$$

To calculate the magnitude of the net displacement, use the Pythagorean theorem as,

$$\left|\vec{C}\right| = \sqrt{A^2 + B^2} \quad (1.7)$$

And to find the direction of the net displacement, use geometry as,

$$\theta = \tan^{-1}\left(\frac{B}{A}\right) \quad (1.8)$$

Example 1.5. What are the magnitudes and directions of the vectors given in figure 1.5?

Solution:

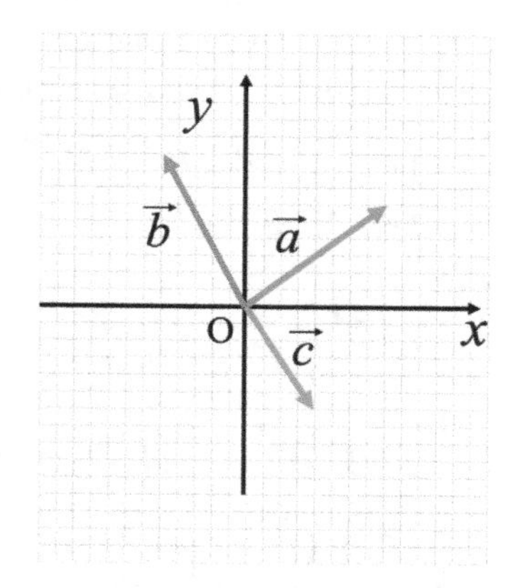

Example 1.6. Given the following vectors: $\vec{A} = 3\hat{i} + 2\hat{j}$ and $\vec{B} = 3\hat{i} - 4\hat{j}$.

(a) Find, $\vec{A} + \vec{B}$

Solution:

(b) Find, $\vec{A} - \vec{B}$

Solution:

Example 1.7. A minivan travels due East on a level road for 32 km. Then turns due North at an intersection and travels 47 km before stopping. Find the vector that indicates the resulting location of the car.

Solution:

1.5.7 Activity-to-do: understanding vectors and its components

Materials needed:

Protractor; meter stick; tape (or any temporary sticky material) to mark distances; open space, paper, calculator.

Experiment-to-do:

1. In your teams, walk to the atrium with a meter stick, a protractor, piece of paper and a calculator.
2. Choose one person to stand near the center facing your choice of corner. Mark this spot with a tape, call it point 0. Make sure you are off-center in the room.
3. Draw that corner on your paper and label the two walls you are facing as Wall 1 and Wall 2.
4. Measure the perpendicular distances from wall 1 and wall 2 and also measure the diagonal distance from the corner. Call these distances $D1$, $D2$, and D respectively. Make sure to use SI units.
5. Your paper should look similar to figure 1.6.
6. Repeat the above procedure moving diagonally towards the corner and record your measurements in table 1.2. Hint: keep equal steps.
7. Next, starting from the 0 point, take steps in equal increments and record your measurements in the table.
8. Use this value of (θ) to finish the rest of the table below: (hint: you will need to use a calculator to find the values of cos(θ) and sin(θ)).
9. Are there some distances that look the same? Can you explain why?

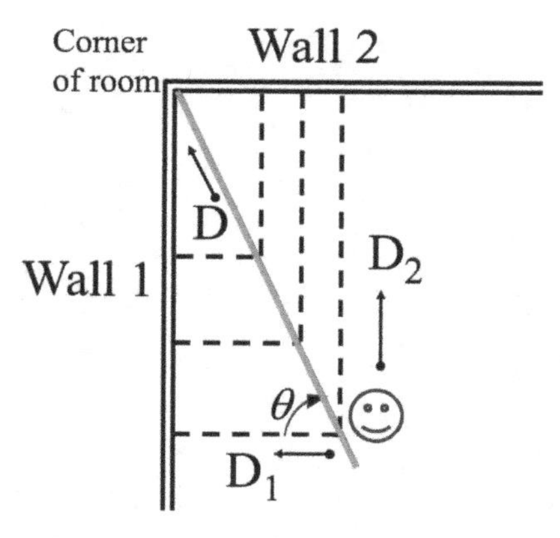

Figure 1.6. Understanding vectors and their components.

Table 1.2. Understanding vectors.

	Distances				Components of step towards the wall		
	$D1$	$D2$	$D3$	Angle	a_x	a_y	Vector ($\bar{a}$)
	m	m	m	θ	$a_x = D3\cos\theta$	$a_y = D3\sin\theta$	($\bar{a}$)
Step 0							
Step 1							
Step 2							
Step 3							
Step 4							

1.6 Speed versus velocity

1.6.1 Speed of an object

The speed of an object is the numerical value of the distance traveled by an object divided by the time taken by the object to cover this distance. Speed is also called the rate of change of distance. Speed describes how fast an object is moving and hence it will always be a positive number (scalar). Mathematically, speed is:

$$\text{Speed} = \frac{\text{Distance traveled}}{\text{Time taken}} \tag{1.9}$$

The average speed of an object is the distance traveled by the object divided by the duration of the interval. Mathematically, given as,

$$\text{Average speed} = \frac{\text{Total distance traveled}}{\text{Elapsed time}} \tag{1.10}$$

The instantaneous speed is the value of the average speed in the limit when duration of time interval approaches zero. The speedometer reads the instantaneous speed of your car and is directionally independent.

$$\text{Instantaneous speed} = \lim_{\Delta t \to 0} \frac{\text{Total distance traveled}}{\text{Elapsed time}} \tag{1.11}$$

Example 1.8. While crossing from the US to Canada, you see that the speed limit on the highway is posted to be '90 km h^{-1}'. What is that in miles per hour? (Hint: 1 mile = 1.6 km.)

Solution:

Example 1.9. How far will a car travel in 15 minutes if it is going 20 m s^{-1}?

Solution:

1.6.2 Velocity of an object

The velocity of an object is the 'vector speed' of an object, also called the rate of change of displacement. To define velocity we need to define a direction associated with the movement in addition to the speed of the object. Also, average speed is generally not related to the magnitude of the average velocity. An Indi-500 race car has an average velocity = 0 around a lap!!!! Mathematically, velocity is:

$$\text{Velocity} = \frac{\text{Displacement}}{\text{Time taken}} = \frac{\Delta \vec{r}}{\Delta t} \tag{1.12}$$

The average velocity of an object in any interval is the change in displacement divided by the duration in which the displacement occurs.

$$\text{Average velocity} = \frac{\Delta \vec{r}}{\Delta t} \tag{1.13}$$

The Instantaneous velocity is the value of the average velocity in the limit when the duration of time interval approaches zero. The magnitude of the instantaneous velocity is the speed and the direction of the instantaneous velocity is the direction in which the motion is occurring.

$$\text{Instantaneous velocity} = \lim_{\Delta t \to 0} \frac{\Delta \vec{r}}{\Delta t} = \frac{d\vec{r}}{dt} \tag{1.14}$$

Please note that since velocity is a vector quantity we have to use the vector laws in order to add, subtract, multiple or divide the velocities as is true for all vector quantities.

Example 1.10. A car moved 20 km east and 60 km west in 2 hours.

1. What is its average speed?
2. What is its average velocity?

Solution:

1.7 Graphical representation of motion

One of the most effective methods of describing and understanding motion is by graphing it. We can tell a lot about the motion of a body by simply plotting graphs of distance, velocity and acceleration versus time. Things like how fast an object is moving, how far it has moved and whether it is slowing down or speeding up can be determined from the said graphs.

Various common situations we encounter in life are discussed below using graphs.

1. **Object at rest:** when a particle is at rest, its position does not change with respect to time. The particle remains at coordinate 'A' at all times. Mathematically, expressed as,

$$x(t) = \mathrm{A} = \mathrm{m}.$$

2. **Object with a constant speed:** distance increases linearly with time while velocity remains constant. This is like a car driving down a highway with cruise-control turned on at 55 mph. The car will maintain its speed at 55 m h^{-1} and will be 55 miles away one hour later.

$$\text{Speed} = \frac{\text{Distance}}{\text{Time interval}} = \frac{\Delta x}{\Delta t} = \frac{\mathrm{m}}{\mathrm{s}}.$$

3. **Object with an increasing velocity:** an object increasing in velocity is said to be undergoing an accelerated motion. The acceleration can also be called the rate of change of velocity. The SI unit for acceleration is meter per second squared (m s^{-2}).

$$\text{Acceleration} = \frac{\text{Velocity}}{\text{Time interval}} = \frac{\Delta \overline{v}}{\Delta t} = \frac{\mathrm{m}}{\mathrm{s}^2}.$$

An object decreasing in velocity is said to be undergoing a decelerated motion. Deceleration is expressed mathematically by adding a negative sign to the value. This is why acceleration is a vector quantity.

1.7.1 Activity-to-do: understanding motion with a motion detector

Materials needed:

Logger-pro software; motion detector; meter stick.

To do:

Look at the following graphs and try to explain the motion of the object that is making the graphs. Write down your hypothesis regarding the motion in the space provided below. Next, try to duplicate them yourself with the help of a motion detector.

1.

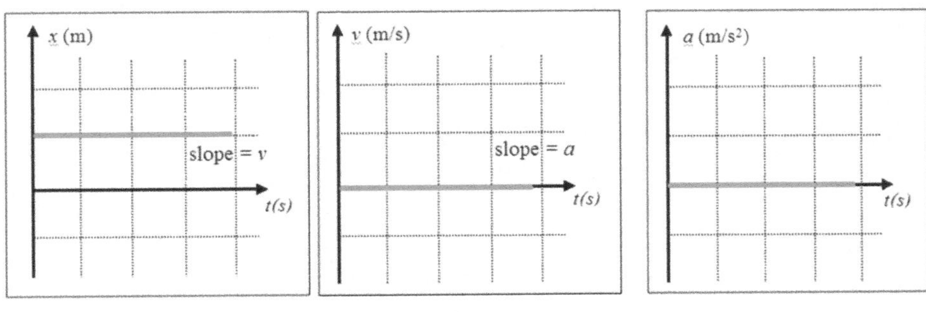

Hypothesis:

Duplicated by doing the following ....

2.

2(a) 2(b) 2(c)

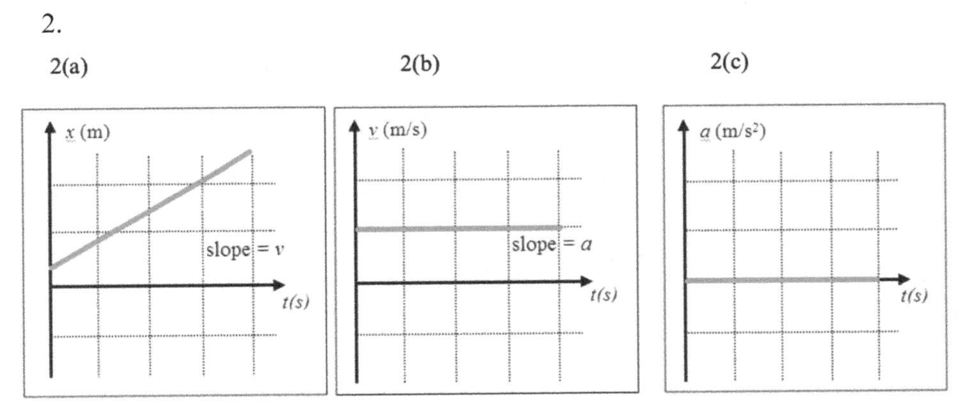

Hypothesis:

Duplicated by doing the following ....

3.

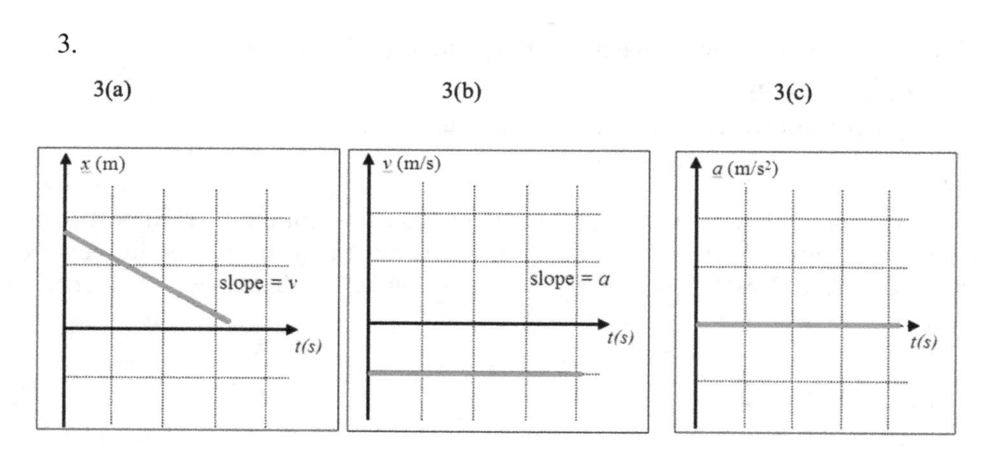

Hypothesis:

Duplicated by doing the following ....

4.

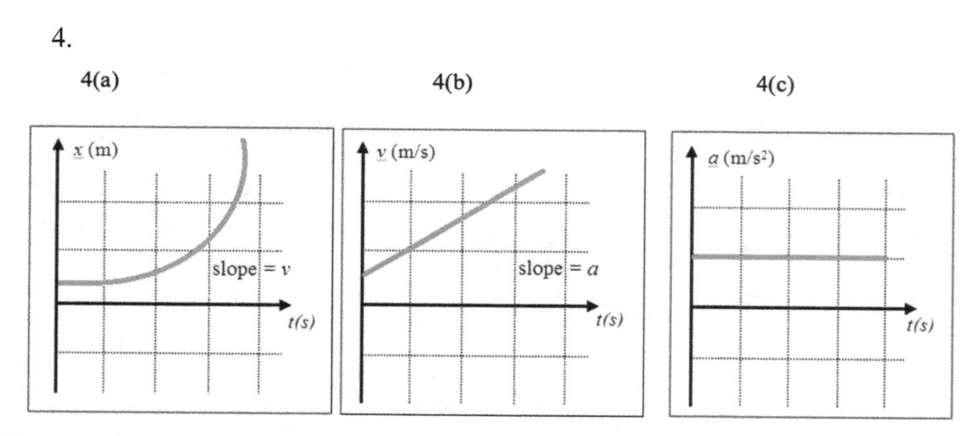

Hypothesis:

Duplicated by doing the following ....

5.

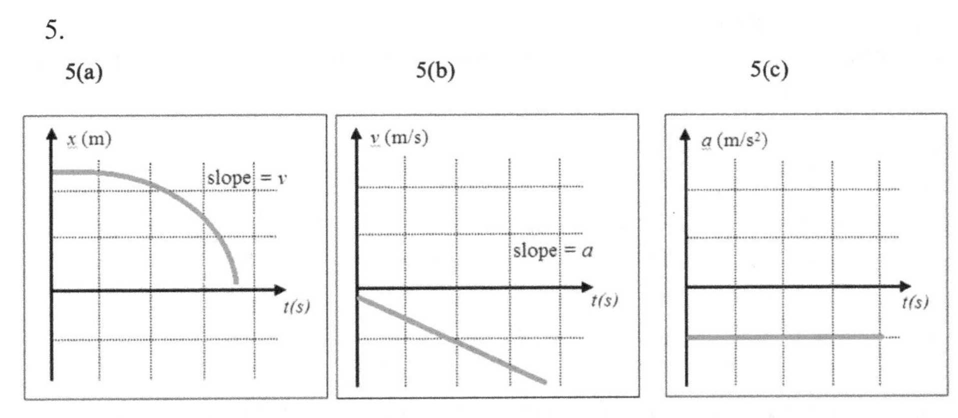

Hypothesis:

Duplicated by doing the following ....

6.

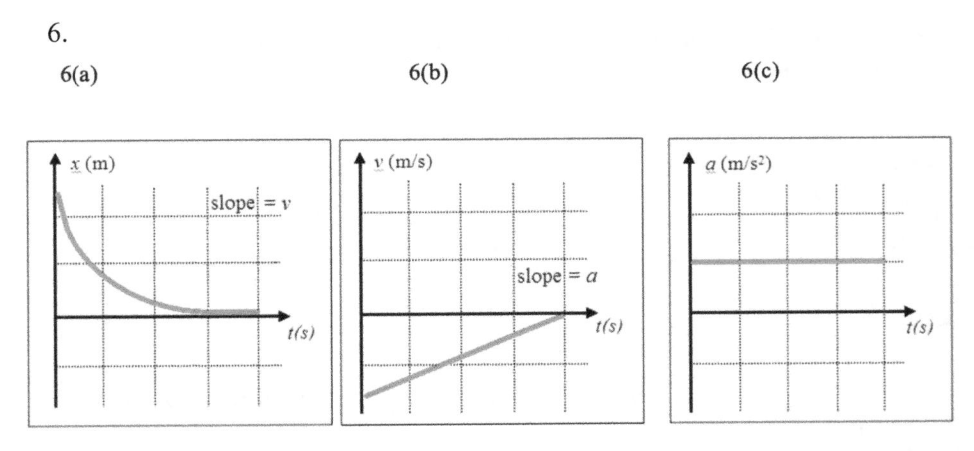

Hypothesis:

Duplicated by doing the following ....

Discussion questions

1. What type of motion occurs when the slope of the distance versus time graph is constant?

2. What kind of motion is occurring when the slope of the velocity versus time graph is zero?

3. What kind of motion is occurring when the slope of the velocity versus time graph is not zero?

4. Draw the distance, velocity and acceleration versus time graph of an object at rest.

Chapter 2

Sound, music and noise

2.1 What to know about sound, music or noise

1. Acoustics
2. Sound
3. Music
4. Noise
5. Branches of acoustics
6. Sound waves
7. Wave detectors—eyes, ears and touch
8. Mechanical waves
9. Electromagnetic waves
10. Longitudinal waves
11. Transverse waves
12. Sound wave propagation—compressions and rarefactions
13. Sound needs—source, vibrating medium, reciever
14. Radiating patterns for sound.

doi:10.1088/978-0-7503-6350-1ch2

2.1.0.1 Activity-to-do: longitudinal versus transverse waves

Materials needed:

Slinky, clamps.

To do:

One way to visualize the movement of longitudinal and transverse waves is by using a slinky toy. Set up the slinky such that it is stretched between its two ends.

1. **Longitudinal waves:** if you push the slinky along its length such that the coils of the slinky are compressed, you will observe the compression traveling down the length of the slinky (figure 2.1.).
2. **Transverse waves:** to set up a transverse wave, position a slinky toy again stretched between two ends. If the slinky is jiggled sideways or to and fro perpendicular to its length, you will observe a motion that travels down the length of the slinky (figure 2.2).

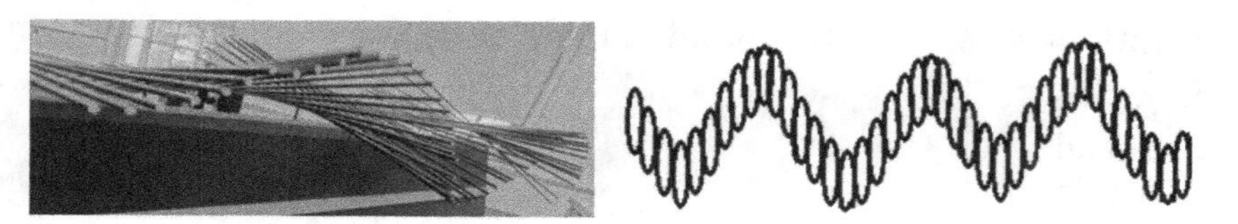

Figure 2.1. Transverse waves, wave propagates to the right, whereas the medium moves perpendicular to the motion of the wave.

Figure 2.2. Longitudinal waves, regions of compressions are depicted by black, whereas rarefactions are shown in gray.

2.1.0.2 Activity-to-do: longitudinal versus transverse waves

Materials needed:

Many friends; a parcel that you can pass between you.

To do:

Stand in a line and try to see how you would make transverse and longitudinal waves.

2.1.0.3 Activity-to-do: feeling vibrations through mediums

Materials needed:

Three balloons that are inflated at different capacities/volumes.

Experiment-to-do:

1. Start with three balloons inflated to different volumes,
   (a) balloon one inflated to the max,
   (b) balloon two inflated to medium capacity, and
   (c) balloon is left to be squishy, as shown in figure 2.3 (left).
2. Hold the balloon with both hands.
3. Use one hand (right hand) to create vibrations at one side of the balloon as shown in figure 2.1 (right), by scraping your right hand on the balloon.
4. Repeat the motion for all three balloons.

1. **Observation I**

   What does your left hand feel as your right hand scrapes the right side of the balloon?

2. **Observation II**

   As your hand scrapes the balloons, explain in your own words what you hear. Hint: if you are familiar with the term pitch, please explain what you observe and the different pitches that are set up as the result of your hand-motion on the three balloons.

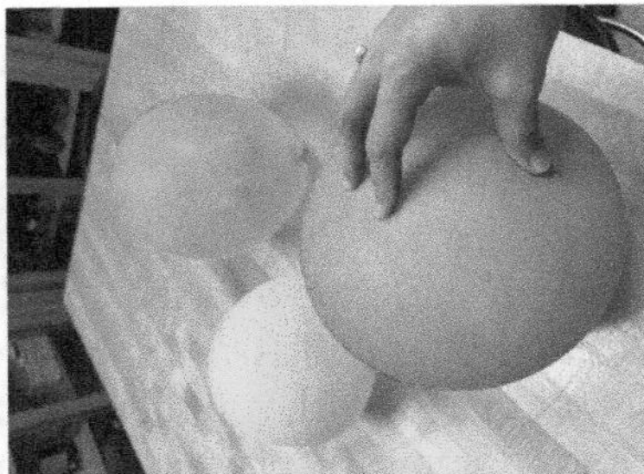

Figure 2.3. (Left) inflate balloons to different capacities. (Right) use one hand to create vibrations by scraping.

2.1.0.4 Activity-to-do: making vibrations and hearing pitches with plastic rulers

Materials needed: Rulers (flexible and stiff), tape.

Experiment-to-do:

1. Hold rulers to find the resonance points as seen in figure 2.4 (left).
2. Once you have found the resonance point, tape the ruler to the table and make music as seen in figure 2.4 (right).

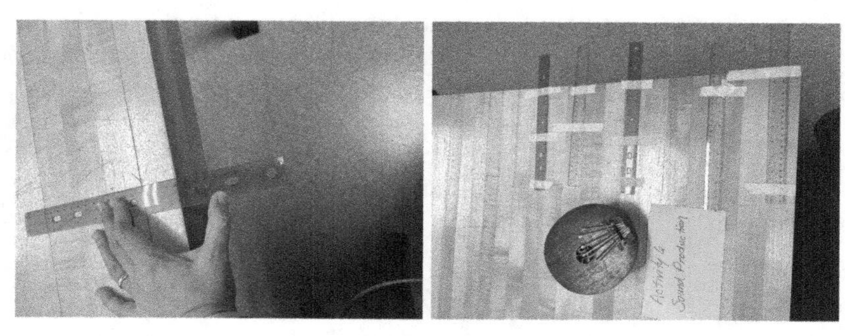

Figure 2.4. (Left) hold rulers to find the resonance points. (Right) tape the rulers to the table and make music.

2.1.0.5 Activity-to-do: mysterious magic—vibrating source

Let's start our investigation of sound by making some salt/sprinkles move without touching them. Is this phenomenon a mystery? Is it magic, possibly telepathy? Or is it science?

Materials needed:

One rubber band; one piece of plastic; one large can; one wood ruler; one small can; salt (figure 2.5).

Experiment-to-do:

1. Find a place to work that is not too close to another team.
2. Take both the large and the small cans.
3. Pull the plastic tightly over the open end of the large can and hold it while your partner puts the rubber band over it.
4. Sprinkle some salt/sprinkles on top of the plastic on the large can.
5. Hold the small can (upside down) close to the salt/sprinkles on the large can.
6. Tap the side of the small can with a ruler. What happens to the salt/sprinkles on the large can?

Figure 2.5. Mysterious magic.

7. Try tapping the small can in different spots or holding it in different positions and directions. What is the best combination to get the salt/sprinkles to move. Write your observations here.

1. **Observation I**

   See the salt/sprinkles bouncing up and down? This phenomenon occurs because the plastic is vibrating. The plastic is vibrating because of the sound waves hitting it. Sound is vibrations that travel through the air.

2. **Observation II**

   Can you get the salt/sprinkles bouncing without touching the membrane by another means? (Hint: look for tuning forks in the lab!)

2.1.0.6 Activity-to-do: sound transmission—cup phones

Let's use what we've learned to make a telephone.

Materials needed: Two cups (assorted materials, say paper, plastic, foam etc); two paper clips; one long string (assorted materials, say cotton, plastic, metal etc) (figure 2.6).

Figure 2.6. An example of a cup phone.

Experiment-to-do:

1. Put one end of the string through each cup and tie one paper clip to the end of the string in the each of the cups.
2. Hold one cup by your mouth while your partner holds the other by his or her ear. Move away from each other until the string is taut.

Discussion questions:

1. Leave a lag in the string. Is it easier or harder for your partner to hear when the string has a lag in it?

2. Have your team member face away from you. Say 'good morning' without and with using the cup phone. Is it easier or harder for your partner to hear you with the cup phone?

3. Try cups composed of different materials. Which material was the best conductor of sound?

4. Try strings composed of different materials. Which material did you find to be the best conductor of sound?

5. Why isn't your family telephone made out of string and cups?

2.1.0.7 Activity-to-do: how fast does sound travel in air?

Materials needed: Stopwatch, a noise maker (drum and drum stick); thermometer; paper; marking tape; meter stick.

Experiment-to-do:

1. Measure 100 m in a straight line on kurtz lane, mark one side as O.
2. Record outdoor temp in °C.
3. How long do you think it will take for the sound to travel 100 m?
4. Stand on one side, and send two friends to the other side.
5. One friend faces you (watcher) and the other away from you (listener).
6. Make one strike on the drum, the watcher starts the stopwatch when they see you strike.
7. The listener hears the sound and says 'stop' as soon as they hear it. The watcher stops as soon as they hear 'stop'.
8. Record the time at least to the 1/10th of a second, this will be the time interval (Δt).
9. Repeat 4–5 times and record your results.
10. Switch roles and repeat.
11. Use the formula, to calculate the speed of sound;

$$\text{Speed} = \frac{\text{Distance}}{\text{Time interval}} = \frac{\Delta x}{\Delta t} = \frac{100\text{m}}{\Delta t} = \frac{\text{m}}{\text{s}}.$$

IOP Publishing

The Physics of Sound and Music, Volume 2
A complete course text (Lab manual)
Samya Bano Zain

Chapter 3

Music, History and Culture

3.1 Activity-to-do: create a timeline of musical eras

Materials needed:
paper; pen.

Experiment-to-do:
Create a timeline of musical eras; for full credit you must include at least eight eras. Examples of timelines.

1. Puzzle timeline.
2. Storyboard a timeline
3. Colorful road map timeline.
4. Chain-link timeline.
5. Clothes line timeline.

doi:10.1088/978-0-7503-6350-1ch3

3.2 Activity-to-do: place the musical eras on the world map

Materials needed:

Google Earth; google maps; paper; pen.

Experiment-to-do:

1. Open Google Earth or Google Maps on your devices.
2. Working in a team, pick a particular musical era you found in activity 3.1.
3. Discover the history, geography, street view of places that correspond to your chosen musical era.
4. Discuss within your group the details you find.
5. Make a PowerPoint slide.
6. Detail why the era you picked is the best and upload PowerPoint to canvas opinion.

Part II

Sound production

IOP Publishing

The Physics of Sound and Music, Volume 2
A complete course text (Lab manual)
Samya Bano Zain

Chapter 4

Tension and deformations in a string

'*May the force be with you!*'

4.1 What to know about energy and force

1. Kinematics versus dynamics
2. Force
3. Energy
4. Kinetic and potential energies
5. Conservation of energy
6. Force and acceleration
7. Force and deformation
8. Elasticity versus hardness
9. Newton's Laws of motion
10. Normal force
11. Friction—static versus kinetic
12. Impact
13. Collision
14. Tension
15. Weight
16. Mass versus weight.

Kinematics, is the branch of physics that describes the motion of objects in space without considering the cause of the motion, whereas the branch of physics that deals with energy and force is called '*dynamics*'.

doi:10.1088/978-0-7503-6350-1ch4

4.2 Activity-to-do: conservation of energy—ball drop

Materials needed:

Three different types of balls, (e.g. tennis ball, golf ball, pingpong ball, bouncy rubber ball etc); meter stick; video recorder (cell phone).

Background information:

Force in physics is described as either a 'push' or a 'pull'. Forces have both a magnitude and a direction, thus they are described using vectors. Energy (E) is defined as the ability of an object to do work. In physics, work is said to have been done on an object when an applied force ($\vec{F}$) moves it over a certain displacement ($\vec{x}$).

$$\text{Work (W)} = \vec{F} \cdot \vec{x} \tag{4.1}$$

This means, when we want an object to do more work we need to provide it with more energy. Both energy and work have the same units, called '**joules**' (J).

One of the most important laws in physics is the law of conservation of energy. Law of conservation of energy states that for closed systems[1] energy cannot be created or destroyed, however, it can be converted from one form to another. Mathematically, conservation of energy is written as,

$$\Delta E = E_{\rm f} - E_{\rm i} = 0 \tag{4.2}$$

where $E_{\rm f}$ represents the total final energy of the system and $E_{\rm i}$ represents the total initial energy of system.

Conservation of energy means that energy cannot be created or destroyed. However, one form can be converted to another form, for example, kinetic energy ($\text{KE} = \frac{1}{2}mv^2$) can be converted into potential energy ($\text{PE} = mgh$) and potential energy can be converted into kinetic energy. For example, when you rub your hands together on a cold day you are converting kinetic energy due to the motion of your hands into heat energy that warms you. In fact, what may seem like a 'loss of energy' in a system is actually energy being converted into heat or sound or both.

Experiment-to-do:

This activity shows how energy converts from one form to another and how loss of energy occurs in a collision.

1. Experiment with the various provided balls.
2. Drop each ball from the 50 cm and the 1 m mark on the meter stick for at least three trials.
3. Record the drop, from the initial height until the rebound height of the first bounce.
4. Calculate both initial height and the average rebound height by reviewing the movies in tables 4.1 and 4.2.

[1] A closed system is a physical system that does not allow transfer of mass or transfer of energy across its boundaries.

Table 4.1. Ball drop activity—drop height = 0.5 m.

	Type of ball	Trial 1	Trial 2	Trial 3	Average rebound height	Average difference b/w rebound and drop
Ball 1						
Ball 2						
Ball 3						

Table 4.2. Ball drop activity—drop height = 1.0 m.

	Type of Ball	Trial 1	Trial 2	Trial 3	Average rebound height	Average difference b/w rebound and drop
Ball 1						
Ball 2						
Ball 3						

Discussion questions:

1. Name the kind of potential energy that this system will have and write the mathematical formula for it.

2. What was the initial potential energy of each ball at 0.50 m and at 1.0 m?

3. In terms of energies, explain what occurs when a ball drops from a height.

4. Did all balls rebound to the height they were dropped from?

5. Which ball bounced the highest?

6. What is the difference between the initial potential energy and the return energy?

7. Why is each ball not able to bounce back to its original drop height?

8. What change occurs in the energy of a falling ball after it bounces?

4.3 Activity-to-do: deformation in collisions

Materials needed:
Modeling clay (Plasticine) rolled into a ball; slow-motion camera.

Background information:
There are two basic effects of an applied force on an object,

- **Force and acceleration:** when an unbalanced force acts on an object it causes an acceleration ($\vec{a}$) or deceleration in the object in the direction of applied force. Newton's second law states,

$$\Sigma\vec{F} = m\vec{a} \tag{4.3}$$

  where m is the mass of the object, measured in kilograms (kg).
- **Force and deformation:** another way that an applied force effects an object is by physically changing the shape of the object. A change in shape of an object due to the application of a force is called a '*deformation*'.
  - **Elastic deformation**: elastic deformation occurs when the deformation is small and the object returns to its original shape once the force (called stress) is removed. Under elastic deformation, size of the deformation is proportional to the force and Hooke's law is obeyed, given by,

$$F = k\Delta L, \tag{4.4}$$

    where ΔL is the change in length and k is the spring constant.
  - **Inelastic deformation**: under inelastic deformation, the object does not return to its original shape once the force is removed.

Experiment-to-do:
When objects collide with each other, some of the initial kinetic energy just before impact is transformed into sound energy, and heat from friction (thermal energy), while some of it becomes elastic potential energy resulting from the instantaneous deformation of the object. In this experiment we will see the impact this has on the shape of the colliding object.

1. Roll the modeling clay into a ball.
2. Drop the modeling clay ball from a height of 1 m and 2 m.
3. Record the impact point in slow motion.
4. Re-watch the video before/during/after the point of impact.
5. What do you observe?

Discussion questions:

1. What affect did dropping the modeling clay ball from 1 m versus 2 m have on the impact?

2. Was the ball deformed more when dropped from 1 m or when it was dropped from 2 m?

4.4 Activity-to-do: collision impact

Materials needed:
Videos recorded in 'Conservation of energy—ball drop' activity 4.2.

Background information:
When the ball collides with the ground, some kinetic energy is transformed into sound energy and heat from friction (thermal energy) and some of it becomes elastic potential energy resulting from the instantaneous deformation of the ball when it collides with the ground. The potential energy of a compressed spring is,

$$\mathrm{PE} = \frac{1}{2}kx^2 \tag{4.5}$$

where k is the spring constant and x is the amount of compression of the spring.

How much any particular material resists a change in its shape is called the 'hardness' of the material. Hard materials resist pressure, are usually more brittle, and are more difficult to cut and to shape. This means that hard materials tend to shatter rather than bending. The Mohs scale of mineral hardness is a qualitative ordinal scale, from 1 to 10. Diamond, an allotrope of carbon, is the hardest substance found on Earth, it ranks as a 10 on the Mohs scale, whereas elastic materials, like rubber, are easily deformable and not strong at all. However, elastic materials can hold a lot of energy.

Experiment-to-do:

1. Watch the videos recorded during activity 'Conservation of energy—ball drop' for the ball dropped from 1 m.
2. Zoom in to the point in the video that shows the different balls hitting the ground.
3. If you do not have the videos, please re-record the videos for the ball dropped from 1 m, focussing in at the impact point.
4. Watch the videos in slow motion.
5. What do you observe for each ball?

Discussion questions:

1. Explain in your own words what you observe.

2. Did the shape of some balls change more than others?

3. What do you note about the balls whose shape changed more upon impact than others? Think elasticity versus hardness of materials.

4. What energies did the initial kinetic energy transfer to?

4.5 Activity-to-do: mass versus weight

Materials needed:

Force gauge; multiple small masses (say M&Ms or Skittles candy); balance scale; weight hanger, clamps.

Background information:

In normal conversations terms mass and weight may be used interchangeably, however, weight and mass in physics are not the same thing.

1. **Mass:** the property of the object that affects our ability to accelerate it is called its 'mass'. Hence, mass is '*the property of a body that that resists the change in its motion.*' Mass is a scalar quantity (basically a number), its units are kilograms and it is independent of the physical location of the object. Another definition of mass is '*the quantity of matter inside the object*'.
2. **Weight:** weight is a force and is due to the force of gravity attracting one object towards another. The SI unit of weight is newtons.

$$\vec{F} = W = mg \tag{4.6}$$

where g is the acceleration due to gravity.

Example 4.1. Which is heavier, 1 kg of feathers or 1 kg of stones?

Solution:

Experiment-to-do:

1. Hang the mass with a force gauge.
2. Measure the weight of various amounts of small objects, staring from one M&M or Skittles candy.
3. Record your results in table 4.3.
4. What do you observe?

Table 4.3. Weight versus mass.

	Number on force gauge	Number on mass scale
1.		
2.		
3.		
4.		
5.		

5. Next, use the same amounts of small objects, staring from one M&M or Skittles candy.
6. What do you observe?

7. Are both numbers the same? If not, why not?

8. Make a graph of weight versus mass, what does the slope of this graph represent?

Example 4.2. Consider a box resting on a table as seen in figure 4.1.

1. Write the force equation for the *x*-component of force.
   Solution:

2. Write the force equation for the *y*-component of force.
   Solution:

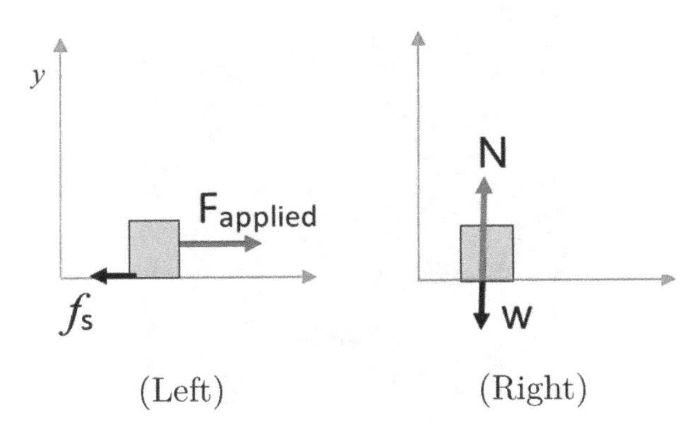

Figure 4.1. A box resting on a table; forces along the *x*- and *y*-axes.

4.6 Activity-to-do: tension and deformation in rubber bands

Materials needed:

One or two rubber bands; measuring ruler (or a meter stick); various masses (50–500 gms).

Background information:

Tension (T), measured in newtons is a pull force exerted by a string, cable, or a chain on an object of mass (m). Suppose a mass is hung on the end of a rope and the rope is suspended from the ceiling by a hook. Tensional force is always directed away from the objects to which it applies and that for our situations the string is always assumed to have negligible mass. In this case, Newton's second law on the mass (m) is,

$$\Sigma F_y = T - mg \tag{4.7}$$

There are two basic possibilities for systems of objects held by strings:

1. **When acceleration is zero:** when the acceleration in the system is zero then equation (4.7) will be reduced to, $T = mg$. The system is in equilibrium.
2. **When the system is accelerated:** when there is acceleration in the system, for example, in the case of Atwood's machine, equation (4.7) will become, $T = ma + mg$.

Experiment-to-do:

1. Cut the rubber band into a long length.
2. Hang one side of the rubber band from a stand and hang a mass hanger from the other side, as shown in figure 4.2.

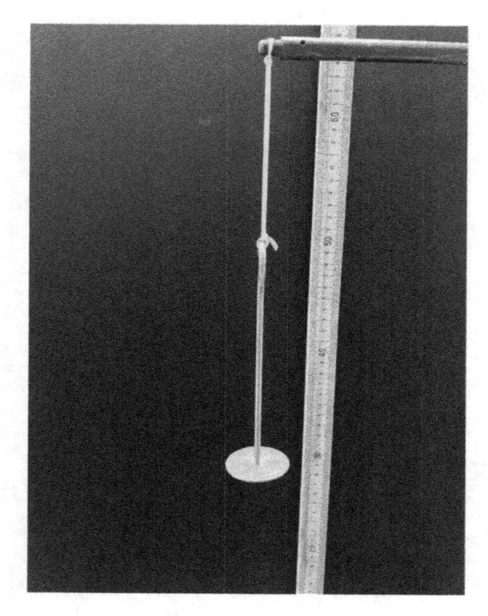

Figure 4.2. Tension and deformation in rubber bands.

3. Use the mass hangers to add masses to one of the rubber bands.
4. Start from mass (m_1) = 50 gms and take a reading in equal increments of increasing mass.
5. Put values in a table 4.4.
6. As you increase the mass on the rubber band, pluck the rubber band and note the change in pitch. Does it become higher or lower with increasing mass? Hint: you may have to hold the masses so they do not fly off the hanger.

Table 4.4. Tension and deformation in rubber bands.

	Mass on band (gms)	Mass on band (kg)	Actual position on ruler (m)	Δy (cm)	Δy (m)	Deformation/ tension ($T = mg$)	Elastic constant (k) $k = \frac{mg}{\Delta y}$
1							
2							
3							
4							
5							
6							

Discussion questions:

1. What do you observe as you increase the mass hanging from the rubber band?

2. What is being measured by the stretch of the elastic string?

3. What is the SI unit of tension?

4. What is the difference between mass and weight? What are the SI units for each?

5. Did the pitch become higher or lower when you increased the mass on the rubber band? Explain in your own words what you think may be happening?

4.7 Activity-to-do: calculate tension (T) in a guitar string

Background information:

The string in a stringed instrument is fixed at two ends and held under a tension (T). To tune a string to a required pitch, you apply tension by turning the tuning pegs, shown in figure 4.3, till the string reaches the required frequency. The factors that influence the frequency or pitch of a guitar string include the string tension, makeup of the string or the string material, thickness or the diameter of the string and the length over which the string vibrates. The string tension in guitar string can be increased by using heavier or thicker gauge guitar strings, or by using a larger guitar or by tuning the guitar a note or half a note up, whereas the string tension in a guitar string can be decreased by using lighter or thinner gauge string, or transferring your guitar strings to a shorter guitar or by tuning it to a note or half a note down.

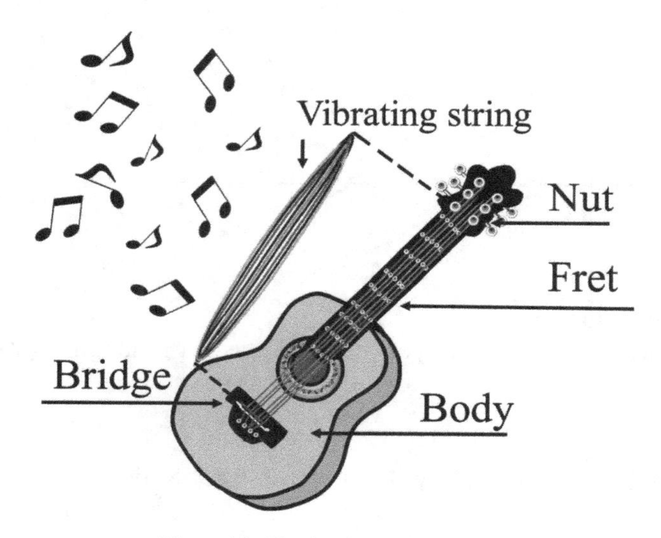

Figure 4.3. Tension in a guitar string.

Example 4.3. Suppose an elevator is held up by a thick rope. Find the tension in the top when:

1. The elevator is moving up with constant velocity.

2. The elevator moves up and accelerates.

3. The elevator moves up but decelerates.

Materials needed:

Various guitar strings; multiple masses; a piece of wood with a pulley to attach the strings; peg; clamps; mass scale.

Experiment-to-do:

To calculate the tension on testing various guitar strings.

1. Set up the experiment similar to figure 4.4.

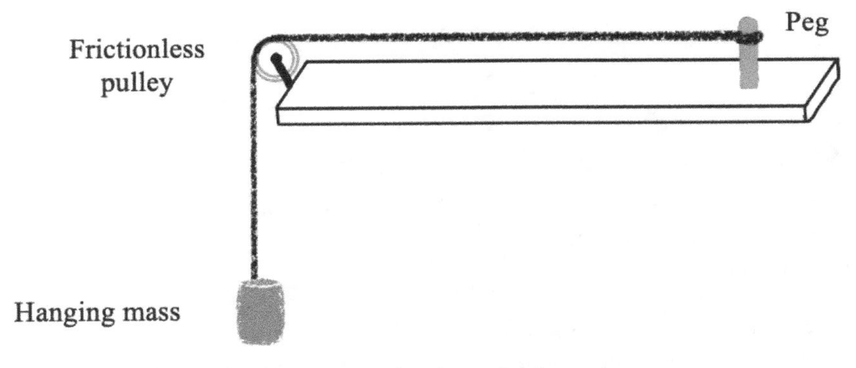

Figure 4.4. Measuring the tension in a string.

2. Calculate mass of hanging portion ($m_{\text{of string}}$), measure the total length of the string (L), mass of the total string (M) and the hanging length (l). Then the mass of the hanging portion of the string is,

$$m_{\text{of string}} = M\frac{l}{L}.$$

Total length of the string (L) = __________

Mass of the total string (M) = __________

The hanging length (l) = __________

3. The tension on the string is equal to the weight of the hanging object plus the weight of the string that is hanging from the pulley,

$$T = \left(m_{\text{on string}} + m_{\text{of string}}\right)g.$$

The tension on the string = __________

IOP Publishing

The Physics of Sound and Music, Volume 2
A complete course text (Lab manual)
Samya Bano Zain

Chapter 5

Vibrating systems

5.1 What to know about vibrating systems

1. Simple harmonic motion (SHM); equilibrium position; displacement; restoring force; inertia
2. Periodic motion
3. Hooke's law
4. Simple harmonic oscillator
5. Examples of SHM; pendulum, physical pendulum, torsional pendulum; mass attached to a spring
6. Properties of waves
7. Energy breakdown of an object undergoing SHM
8. Simple harmonic and circular motion
9. Standing waves
10. Properties of standing waves; equilibrium point; amplitude; frequency; wavelength; period; crest, trough, nodes; anti-nodes
11. Speed of standing waves
12. Allowed and forbidden standing wave conditions
13. Reflection of a wave
14. Standing waves on the string; fundamental frequency, harmonics
15. Harmonics, overtones, partials
16. Wave interaction: superposition and interference
17. Types of wave interference: constructive, intermediate, destructive
18. Beats.

doi:10.1088/978-0-7503-6350-1ch5

5.2 Activity-to-do: waves on a string

Materials needed:

Two volunteers; heavy skipping rope or a piece of string (or spring).

Background information:

All waves share some common wave properties, in terms of waves in a string, as seen in figure 5.1, some properties are:

- **Equilibrium point**: at equilibrium point the disturbance in the string is zero.
- **Wavelength**: the distance between a point on a wave and the equivalent point on the next wave is called the 'wavelength' (λ), measured in meters.
- **Amplitude**: the strength or power of a wave signal is called the amplitude and is the maximum displacement of a particle from its equilibrium position. For sound higher amplitudes mean higher volume.
- **Crest**: the crest of a wave is the point with maximum upward displacement from equilibrium.
- **Trough**: the trough of a wave is the point with maximum downward displacement from equilibrium.
- **Period**: the period is the time for a particle to make one complete cycle.
- **Frequency**: the number of times the wavelength occurs in one second is called the frequency, with SI unit Hertz (Hz). The relationship between period (T_P) and frequency (f) is,

$$f = \frac{1}{T_P} \tag{5.1}$$

- **Speed of standing waves:** for a wave, frequency $f = 1/T_P$ and distance is given as a wavelength (λ). Then,

$$\text{Speed of wave} = \frac{\text{Distance}}{\text{time}} = \frac{\lambda}{T_P} = \lambda f \tag{5.2}$$

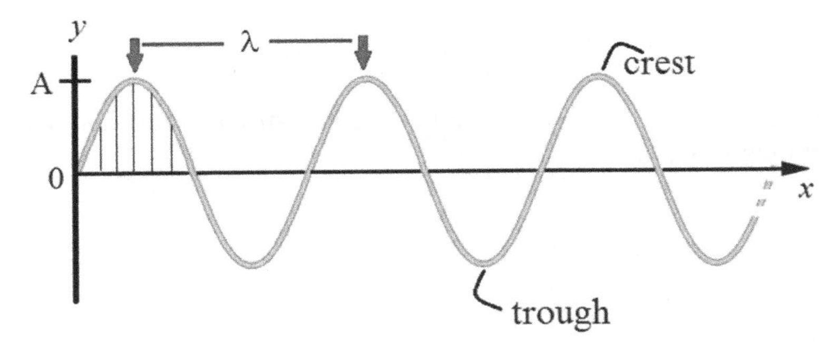

Figure 5.1. A string wave traveling in positive x-axis, 'A' is the amplitude and 'λ' is the wavelength.

Experiment-to-do:

1. Securely hold a string or a heavy spring at both ends between two volunteers.
2. One volunteer should hold their end stationary.
3. Second volunteer please slowly move the string up and down.

4. What do you observe? How many waves were set up?

5. Next, second volunteer should sharply move the string up and down.

6. What do you observe? How many waves were set up?

7. Reverse volunteer roles and try the above experiment again. Do you observe anything different?

8. Change the speed by which you move the string/spring. Do you observe anything different?

5.3 Activity-to-do: standing waves I—string (or spring)

Materials needed:

Two volunteers; heavy skipping rope or a heavy piece of string (or spring); multiple paper clips (colored); two objects to attach the ends of the string to.

Background information:

In a standing wave on a string, as seen in figure 5.2 (left), every point moves in SHM with the same frequency and reaches its maximum distance from equilibrium simultaneously. Standing waves are also called stationary waves.

All standing waves share the common wave properties discussed in activity 5.2. However, In addition standing waves have nodes and anti-nodes. Nodes are points with zero displacements and anti-nodes are places between two nodes at which the displacement oscillates at its maximum at all times. These alternating patterns of the nodes and the anti-nodes together set up a standing wave pattern.

There is no net transfer of energy from one end to the other in a standing wave, and points vibrate with different amplitudes ranging from zero (nodes) to maximum (anti-nodes) amplitude. Since the energy cannot flow past the nodes in the string towards the right or the left, no net energy transfer occurs in a standing wave.

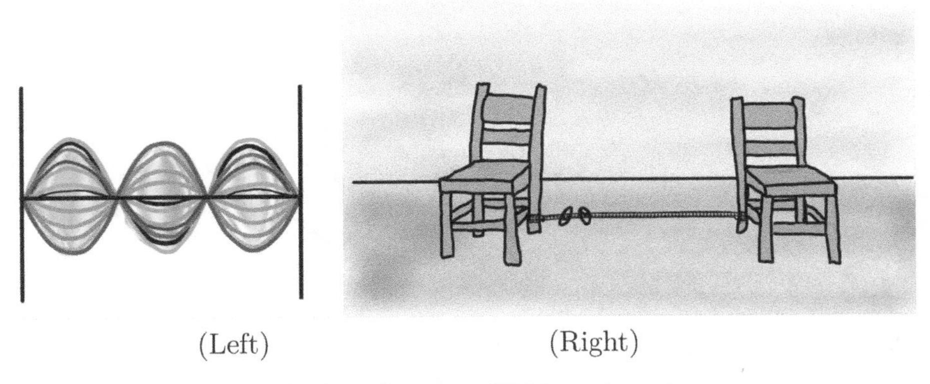

Figure 5.2. (Left) standing waves. (Right) experimental setup.

Example 5.1. What is the fundamental and first harmonic of a standing wave in a string of length 30 cm with speed 20 m s^{-1}.

Solution:

Use the given equation (5.3),

$$f_n = n\frac{v}{2L} \tag{5.3}$$

and know $n = 1$ for fundamental, and $n = 2$ for first harmonic. With given speed of wave as 20 m s^{-1} and length of string $L = 30$ m.

Experiment-to-do:

1. Secure a string between two stationary objects, say classroom chairs, as shown in figure 5.2 (right).
2. Clip paper clips to the string so that they can freely spin around the string.
3. Pluck the string and observe the motion of the paper clips.
4. Change the position of the paper clips.
5. Pluck the string and observe the motion of the paper clips again
6. Change the position where you pluck the string. Can you get other harmonics set up in the string?

5.4 Activity-to-do: standing waves II—nodes in water bottles

Materials needed:
Plastic/glass bottles; water; metal spoon or a wooden ruler (figure 5.3).

Experiment-to-do:
1. Take a bottle and start at the empty level for the water.
2. Gently strike the empty bottle with a spoon or a wooden ruler. What do you hear?

3. Slowly start to pour water in the bottle.
4. Listen carefully. What do you observe about the pitch as the water level rises?

5. Empty the bottle again and start at the empty level for the water.
6. Slowly start to pour in water but this time continuously bang on the slide with a spoon or a wooden ruler.
7. Note how many nodes you can hear in the bottles.

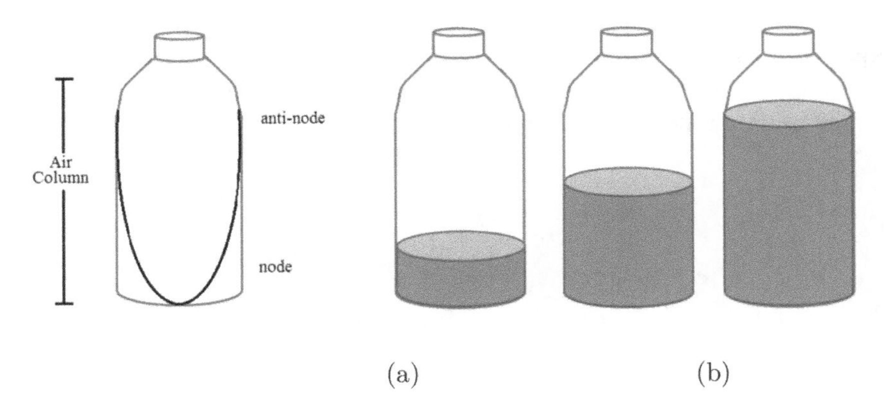

Figure 5.3. (a) Nodes and anti-nodes in water bottles. (b) Water level in bottles.

5.5 Activity-to-do: ‘seeing’ sound waves

Materials needed:

Microphone; oscilloscope or a computer; sound producing device (a human, drum, tuning fork, etc); rubber mallet or another soft surface.

Background information:

As sound waves pass a certain point, the sound pressure rises and falls. Microphones measure this rise and fall. In order to measure the sound waves that pass through a certain point it is useful to make a graph of pressure versus time. This graph is called a ‘*waveform*’ of sound. To display this waveform we connect the microphone to an oscilloscope or a computer, as shown in figure 5.4. A microphone changes the sound waves into an electrical signal. The oscilloscope then shows what these electrical waves look like.

The tuning fork, a U-shaped bar, usually made from steel, was invented in the 1700s by a British musician, John Shore. Tuning forks have traditionally been used to tune musical instruments because they produce a fixed pure tone with no overtones. The pitch of a tuning fork depends on the length and mass of the two prongs.

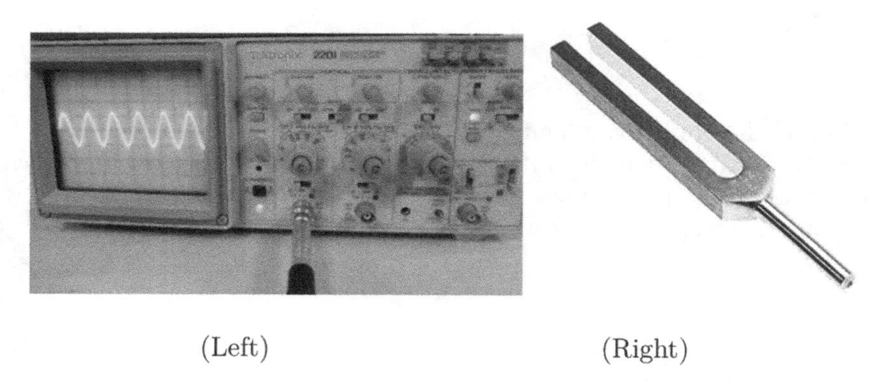

Figure 5.4. ‘Seeing’ sound waves. (Left): oscilloscope. (Right) tuning fork.

Experiment-to-do:

1. Connect the microphone to an oscilloscope or a computer. Sing or talk into the microphone. What do you observe?

2. Take a screen-shot of what you observe or redraw what you observe below.

3. Next use a tuning fork (after striking it on a rubber mallet or another soft surface) into the microphone. What kind of wave do you see? Draw it here.

4. Next whistle into the microphone. What kind of wave do you see? Draw it here.

Discussion questions:

1. What are sine waves?

2. Try to produce a sinusoidal wave on the screen yourself. What did you have to do to reproduce a sinusoidal wave?

3. Draw what you observe below.

5.6 Activity-to-do: SHM—simple pendulum. How does the period depend on the amplitude of the swing in a simple pendulum?

Materials needed:

A pendulum made of a length of a string and a pendulum bob (a mass with a hook); hook or stand to suspend the pendulum; stop watch; protractor; vertical source of light (ceiling light will work well as well); a surface for light to be projected onto (floor or top of table). Note: keep the arc through which the pendulum swings small (20° or less).

Background information:

A common example of periodic motion is the SHM, and the system that oscillates with SHM is called a simple harmonic oscillator (SHO). In SHM the acceleration of the system, hence the net force, is proportional (but opposite in direction) to the displacement. This net force obeys Hooke's law given as,

$$F = kx \tag{5.4}$$

where k is the force constant and x is the displacement away from the equilibrium position.

A pendulum is any object suspended from a fixed point so that it can swing back and forth freely under the influence of gravity alone. This back-and-forth motion is called an 'oscillation.' A simple pendulum is made up of a bob suspended at the end of a string, where the string is light enough with respect to the bob to be considered massless. The formula for the period (T_P) of a pendulum is,

$$T_P = 2\pi\sqrt{\frac{l}{g}} \tag{5.5}$$

where l is the length of the pendulum, measured from the point of suspension to the center of mass of the bob and g is the acceleration due to gravity.

Experiment-to-do:

There is an easy way to produce SHM by using a pendulum and some lights (figure 5.5).

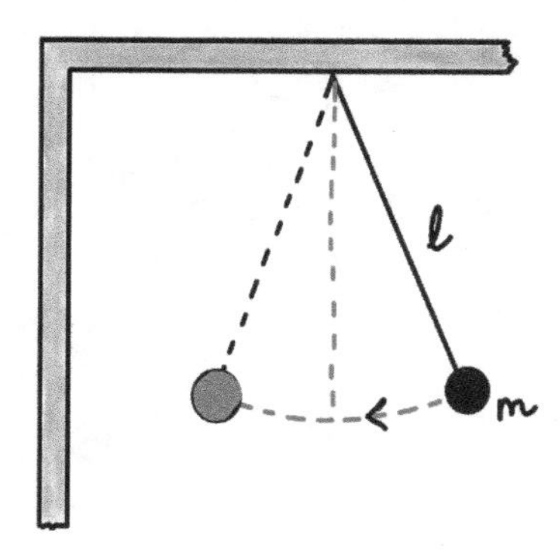

Figure 5.5. Simple pendulum—made of a mass attached to a string.

Table 5.1. Period versus amplitude for a simple pendulum.

	Measured length of string (m)	Mass of bob (kg)	No. of swings	Total measured time (s) Time 1 (t_1)	Time 2 (t_2)	Time 3 (t_3)	Average time (t_{av})	Period (T_P) s
5°								
10°								
15°								
20°								

1. Attach the mass to a string and suspend it from a hook or stand.
2. Project light onto to system from the top.
3. Pull the mass to the right by 5° and let it swing.
4. The movement of the mass will be projected onto the 'projecting surface'.
5. Record the movements of both the mass-string system and the projection for 10 swings.
6. Repeat the process increasing the angle to 10°, 15° and 20°.
7. Find the period of the pendulum by dividing the time for 10 oscillations by 10. Record all values in table 5.1.

Discussion questions:

1. What is the average period of the simple pendulum?

2. What change do you observe in the period as you increase the angle of the string?

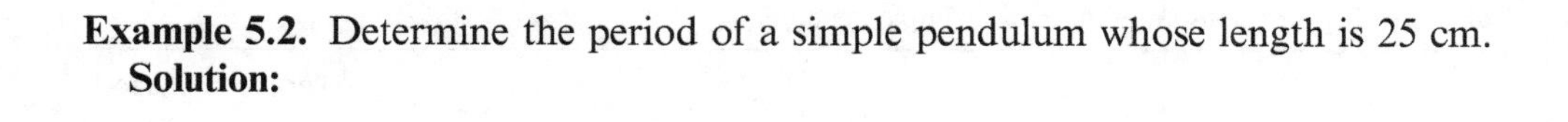

Example 5.2. Determine the period of a simple pendulum whose length is 25 cm.
Solution:

Example 5.3. Find the length of a pendulum which has a period of 7.5 seconds?
Solution:

5.7 Activity-to-do: how does the period depend on the length of a simple pendulum

Materials needed:

A pendulum made of a mass attached to a string; stop watch; protractor; stand. Note: you should always keep the arc through which the pendulum swings small (15° or less).

Experiment-to-do:

1. Attach the mass to a string and suspend it from a hook or stand.
2. In order to measure the period of the pendulum, measure the amount of time for 10 complete swings. To eliminate one source of error, do not use the first couple of swings.
3. Start timing and counting when the pendulum bob returns to its starting point.
4. Repeat this 3 times and average your total times.
5. The period of the pendulum is then the total time measured divided by 10 (i.e., the amount of time for one swing.) The length of the pendulum is the distance from the point of support to the center of mass of the bob.
6. Using the same mass you should measure the period for at least 4 different lengths, ranging from 0.40 m to 1.60 m.
7. All data should be recorded in table 5.2.

Table 5.2. Period versus length in a simple pendulum.

				Total measured time (s)			Average time	Period (T_P)
	Measured length of string (m)	Mass of bob (kg)	No. of swings	Time 1 (t_1)	Time 2 (t_2)	Time 3 (t_3)	(t_{av})	s
1								
2								
3								
4								

Discussion questions:

1. What is the SI unit of period?

2. What do you observe as you increase the length of the string?

3. In order to see more clearly what the data shows, you need to make some graphs in Excel. Plot the period squared of the pendulum as a function of the length. (Hint: the length will be on the x-axis and period squared on the y-axis.)
   (a) Did you obtain a straight line for your plots?

   (b) Can you write an equation for the line?

   (c) What does the slope of this graph represent?

   (d) Calculate the value of acceleration due to gravity (g) from the plot knowing the slope.

   (e) Calculate the % error in your value.

$$\%\text{error} = \frac{(\text{actual } v\text{alue} \;-\; \text{measured average value})}{\text{actual value}} \times 100$$

   (f) How does the period, T_P, depend on length, compare the theoretical prediction and what you obtained experimentally?

5.8 Activity-to-do: how does the period of a pendulum change with length—II

Materials needed:

A mass with a hole through which a string can be threaded; a stick; camera; horizontal source of light; a vertical surface for light to be projected onto, e.g. vertical wall or blackboard.

Experiment-to-do:

1. Thread a mass with a hole with a strong string.
2. Attach the mass to the stick such that it is suspended from both ends of the stick.
3. Have two people hold the stick and move it such that the string wraps around the stick.
4. Project the motion on a screen and record the process until the entire string is wrapped around the stick.
5. Review the video in slow motion.
6. What do you observe?

5.9 Activity-to-do: reflection I: ball and wall

Materials needed:
Tennis ball or any other ball available; protractor; wall; masking tape; meter stick.

Background information:
The bouncing back into the same medium after striking a surface is called 'reflection.' Common examples of reflection include the reflection of light, reflection of sound waves and reflection of water waves.

Laws of reflection:

1. The incident ray, the reflected ray and the normal at the point of incidence lie in the same plane.
2. The angle of incidence (θ_i) is equal to the angle of reflection (θ_r).

Example 5.4. A light ray strikes a reflective plane surface at an angle of 60° with the surface. Calculate:

1. The angle of incidence (θ_i).
   Solution:

2. The angle of reflection (θ_r).
   Solution:

Experiment-to-do:

1. Stand next to a vertical wall, use masking tape to mark a line (at 90°) on the floor as shown in figure 5.6. This line is called the 'normal' and is at 90° from the reflective surface.
2. Sit on the floor and roll a ball (example tennis, golf, basketball etc) toward the wall directly in front of you (on the masking tape). What happens to the ball once it hits the wall?
3. Place a protractor on the floor to measure the angles. One team member may have to hold the protractor off the floor but against the wall for the ball to properly hit the wall.
4. Now, roll the ball (along the floor) at increasing angles to the normal line. You may have to provide a guide for the ball, as seen in figure 5.6.
5. Measure the angle you release the ball from the normal line (angle of Incidence) and measure the angle the ball bounces off the wall (angle of reflection) using the protractor.
6. Tabulate your results in table 5.3.
7. What do you notice about the angle of incidence and the angle of reflection?

Figure 5.6. Reflection—ball and wall.

Table 5.3. Angle of Incidence and angle of reflection for ball on wall.

	Angle of incidence (θ_i)	Angle of reflection (θ_r)
1.		
2.		
3.		

5.10 Activity-to-do: reflection II: pencil and mirror

Materials needed:
Two pencil; mirror; protractor.

(Left)

(Right)

Figure 5.7. Reflection in a mirror, using pencils. (Left) hold pencil 1 as shown here. (Right) move pencil 2 to cover pencil 1 while looking through the mirror.

Experiment-to-do:

1. Stand in front of a mirror facing the mirror and hold a pencil about 20 cm away from the mirror in front of you. What do you observe about the image of the pencil? (Example: how far is the image from the mirror, how far is it from you etc.) Draw a picture if it helps in your explanation.

2. Mark this line using a masking tape (from you to pencil to the mirror) as the normal line. You may also use a thread or a rope if masking tape is not available. The line should be parallel to the surface and perpendicular to the mirror as seen in figure 5.7.

3. Now move the pencil towards the right at increasing angles to the normal line and measure the angle with respect to the mirror, as seen in figure 5.7 (left).

Table 5.4. Angle of Incidence and angle of reflection for pencil and mirror.

	Angle of incidence (θ_i)	Angle of reflection (θ_r)
1.		
2.		
3.		

4. Record this in table 5.4 as angle of incidence (θ_i).
5. On the opposite side of the normal, hold another pencil (pencil 2) such that it covers pencil 1 while looking through the mirror.
6. Read the angle with respect to the normal and record this in table 5.4 as angle of reflection (θ_r).
7. What do you notice about the angle of incidence and the angle of reflection?

5.11 Activity-to-do: reflection III: verify the laws of reflection using lasers

Materials needed:
Laser pointer; ruler; paper; pencil; mirror; protractor.

Experiment-to-do:

1. Put a sheet of white paper on the drawing board.
2. Place a plane mirror vertically on the drawing board.
3. Draw a line $\overline{MM'}$ parallel to the surface of the mirror and mark a point O at the center of the line.
4. Next draw another line normal to the surface and call it $\overline{ON}$ as seen in figure 5.8 (left).
5. Draw $\overline{IO}$, the incident ray in such a way that angle $(I\hat{O}M')$ is less than 90°.
6. Point the laser such that you see the laser light on the paper and it is reflected from the mirror, as seen in figure 5.8 (middle).
7. Stand behind the incident ray, never the reflected ray! (*Don't look directly into the laser, please!*).
8. Draw the reflected image of the laser beam as observed through the mirror on the paper. (*Again, please don't look directly into the laser!*).
9. Measure the angle of incidence (θ_i) and the angle of reflection(θ_r) using a protractor to verify the laws of reflection:
    (a) angle of incidence (θ_I) = angle of reflection (θ_r),
    (b) reflected and incident rays lie in the same plane.
10. Repeat the experiment for multiple angles of incidence.
11. Record this in your data table 5.5.

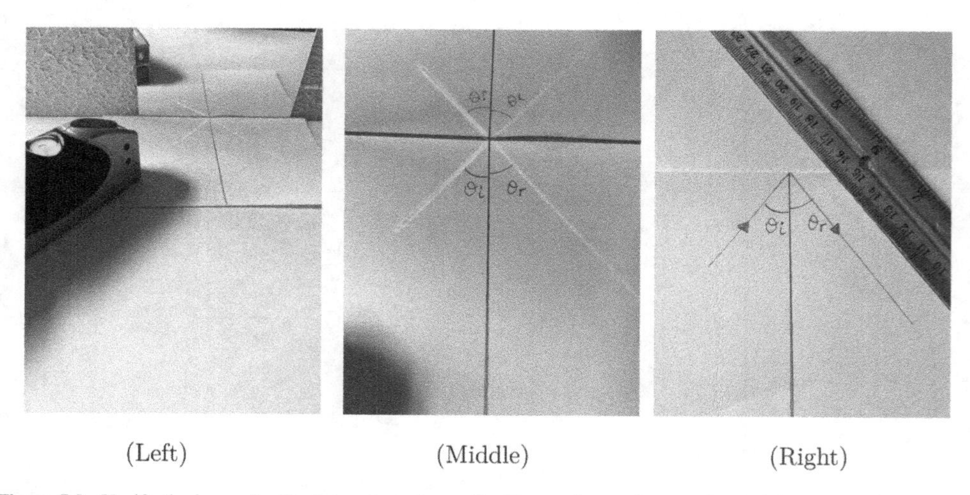

Figure 5.8. Verify the laws of reflection using a laser. (Left) experimental setup; (center) laser turned on; (right) use a ruler to outline the angle of incidence (θ_i) and angle of reflection (θ_r).

Table 5.5. Verify the laws of reflection using lasers.

	Angle of incidence (θ_i)	Angle of reflection (θ_r)	% difference
1.			
2.			
3.			

12. What do you notice about the angle of incidence and the angle of reflection?

5.12 Activity-to-do: reflection IV: verify the laws of reflection using pins

Materials needed: Pins; ruler; paper; pencil; mirror; protractor.

Experiment-to-do:

1. Put a sheet of white paper on the drawing board.
2. Place a plane mirror vertically on the drawing board.
3. Draw a line $\overline{MM'}$ on it and mark a point O at the center of the line and a normal $\overline{ON}$ on $\overline{MM'}$ as shown in figure 5.9.

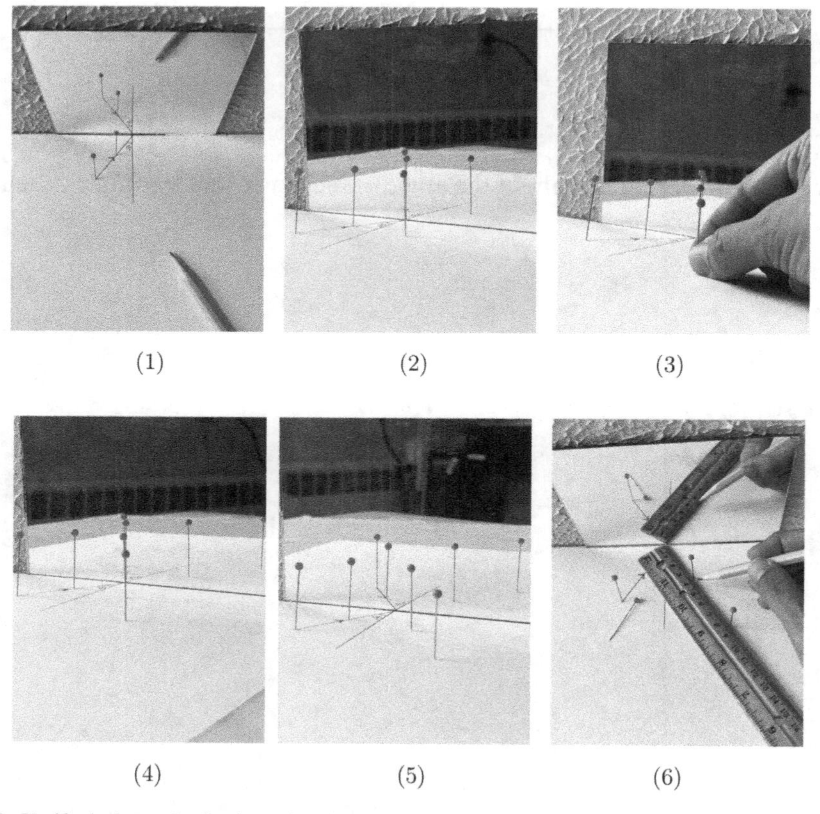

Figure 5.9. Verify the laws of reflection using pins. (1). Draw the normal to the mirror and incident ray, fix two pins (P and Q). (2). Fix pin P′ on the other side of the normal such that it covers the pins (P and Q) while looking through the mirror. (3) Fix pin Q′ such that it covers the pins (P, Q and P′) while looking through the mirror. (4) Side view. (5) Top view. (6) Draw the reflected ray.

4. Draw $\overline{IO}$,the incident ray in such a way that $I\hat{O}M'$ is less than 90°.
5. Fix two pins P and Q on the incident ray $\overline{IO}$.
6. Fix P′ and Q′ on other side of the normal, to represent the reflected image of P and Q, as observed in the mirror as P′ and Q′. Make sure that the

reflection pins (P′ and Q′) completely cover the original/incident pins (P and Q) when seen through the mirror.
7. Circle around all the pins with a pencil.
8. Then remove the pins from the paper and carefully place them back in the box you picked them up from. Please do not leave them on the table.
9. Draw a line $\overline{OR}$ passing through P′ and Q′ to represent the reflected ray.
10. Measure the angle of incidence ($\theta_i = I\hat{O}N$) and the angle of reflection ($\theta_r = N\hat{O}R$).
11. Verify the laws of reflection.
    (a) Angle of incidence (θ_I) = angle of reflection (θ_r),
    (b) Points P, Q, P′, Q′, N and O lie in the same plane.
12. Repeat the experiment for different measures of angle of incidence.
13. Record this in table 5.6.

Table 5.6. Angle of incidence and angle of reflection using pins.

	Angle of incidence (θ_i)	Angle of reflection (θ_r)	% difference
1.			
2.			
3.			

14. What do you notice about the angle of incidence and the angle of reflection?

5.13 Activity-to-do: reflection V: reflections of sound waves

Materials needed:

A slinky, preferably a metal one but a plastic one works just as well; two team members; doorknob or a hook in a wall.

Background information:

Sounds, like all other waves, bounce or get reflected when they encounter an obstacle, such as walls or cliffs, in their path. A reflection of sound that arrives at the listener with a delay after the original sound is called an 'echo'. Humans in general cannot distinguish between the original sound and an echo if the delay is less than 0.1 second. Hence, the reflecting surface must be more than 17.2 m from the original sound for it to be perceived as an echo. The way sound waves are reflected depends on the reflecting surface,

1. **Reflection of sound wave from an open end**: when the incident pulse (compression or a positive pulse) of sound wave approaches an open end, the pressure of the sound wave drops to zero and the pulse reflects back as a rarefaction (or a negative pulse), as seen in figure 5.10(a).
2. **Reflection of sound wave from a closed end**: when the incident pulse of the sound wave approaches a closed end, the pressure of the sound wave builds up and the pulse reflects back again as a compression (or a positive pulse), as seen in figure 5.10(b).

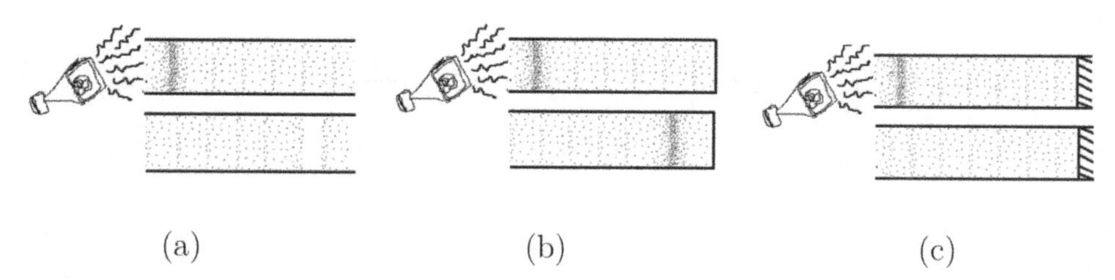

Figure 5.10. Reflection of sound wave from, (a) an open end. (b) From a closed end. (c) From an absorbing end.

3. **Reflection from an absorbing end**: when the incident pulse of sound wave approaches an end with a sound absorber attached to it, the sound absorber will absorb all of the incident pulse and there will be virtually no reflected pulse. An example of an absorbing end is a soft surface, like a pillow. This state is called '*anechoic*' which translates to 'no echos,' as seen in figure 5.10(c).

Experiment-to-do:

1. Attach one end of the slinky to a doorknob or a hook in a wall.
2. Hold the other end and stretch the slinky.
3. With your free hand, gently strike the hand that is holding the slinky, as if you are clapping, as seen in figure 5.11.

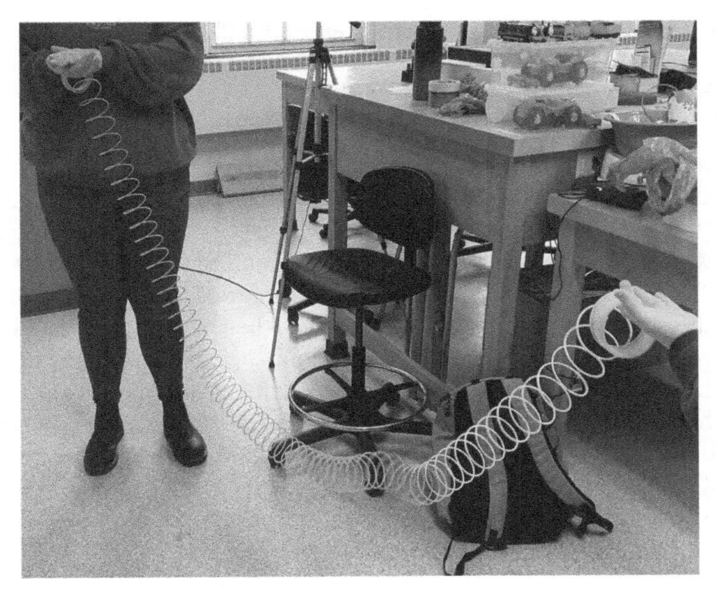

Figure 5.11. Making waves with a simple slinky.

4. Try striking it two or three times in a row and see what happens. What do you observe?

5. Next, have your partner take the slinky off the doorknob and hold it tightly.
6. Again strike your end two or three times in a row. What do you observe?

5.14 Activity-to-do: understanding interference

Materials needed:
PhET simulation interference (wave interference).

Background Information:
When two waves interact with each other at the same time, they 'superimpose' on each other for an instant before continuing on their way. It is interesting to note that the waves will pass through one another and emerge on the other side nearly unchanged. This allows us to distinguish two people speaking in a room at the same time, since the sound waves pass through each other nearly unaffected.

When several waves combine at a single point, the displacement of a particular particle in the medium is a simple linear sum of the individual displacements of the particle. This behavior of waves is called the '*principle of superposition.*' Superposition is the overlap or addition of two waves, whereas interference is the effect of that overlap. In other words, interference is a pattern that you see when you observe a superimposed wave. The interference in water waves is shown in figure 5.12.

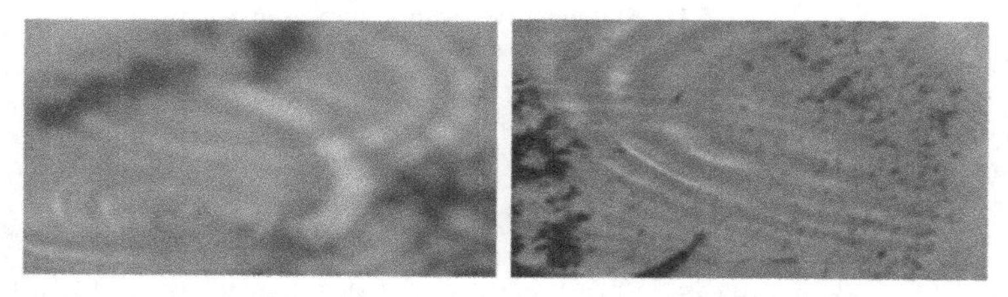

Figure 5.12. Interference in water waves.

Suppose two waves Ψ_1 and Ψ_2 pass through a single point in a medium, say air, at the same time. Then the displacement that a molecule of air feels due to the presence of these two waves will be given as a linear sum $\Psi_{(1+2)}$, mathematically written as,

$$\Psi_{(1+2)} = A_1\Psi_1 + A_2\Psi_2 \tag{5.6}$$

where A_1 and A_2 are the amplitudes of the incoming waves 1 and 2.

Two incoming waves can both either be in-phase or out-of-phase with each other. In-phase means the two waves are doing same thing at same time and there is no phase difference, whereas out-of-phase waves do not do the same thing at same time. There are three ways in which waves can interfere with each other, they can interfere constructively, destructively, or intermediately.

1. **Constructive interference**: constructive interference occurs when both incoming waves are exactly in phase, and such waves add together to produce a stronger wave. The amplitudes of the individual waves add, and the resultant wave is the sum of amplitudes, given as $|A_1 + A_2|$, as shown in figure 5.13(a). In sound waves constructive inference will result in a louder sound.

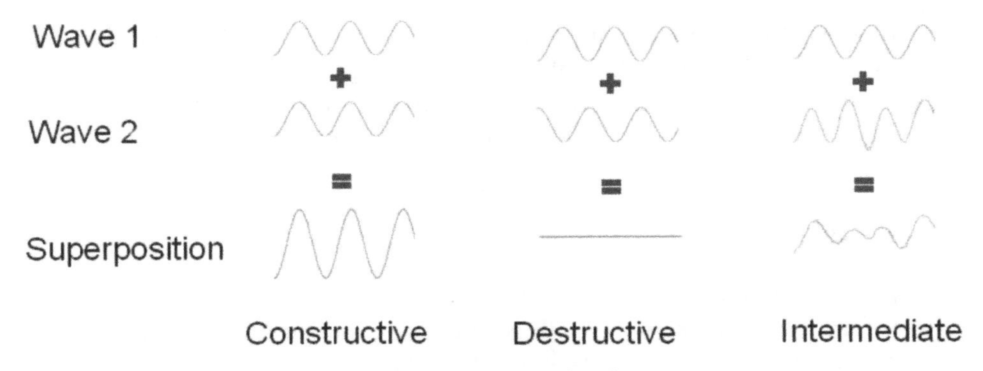

Figure 5.13. Types of wave interference.

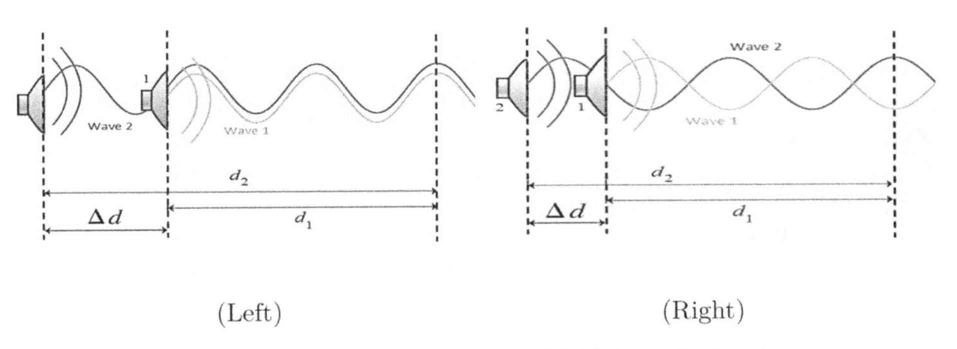

Figure 5.14. (Left) constructive interference. (Right) destructive interference.

2. **Destructive interference**: destructive interference occurs when both incoming waves are exactly inverted, or 180 degrees out of phase, and cancel each other out. The amplitudes of the individual waves subtract; a resultant wave $= |A_1 - A_2| = 0$, as shown in figure 5.13(b). During destructive interference, the energy of the wave is in the form of kinetic energy of the medium. In sound waves destructive inference will result in absolute silence and this is how many noise-cancellation devices work.
3. **Intermediate Interference**: Intermediate Interference occurs when incoming waves have a varying phase relationship. The superposition of these waves has a resultant amplitude somewhere between a maximum of $|A_1 + A_2|$ and a minimum of $|A_1 - A_2|$, given in figure 5.13(c).

Interference of waves from two sources

Suppose two loudspeakers produce identical sound waves, and are placed as shown in the figure 5.14 (left). If d_1 and d_2 are the distances from the loudspeakers to the observer, the path length is the extra distance traveled by the wave coming from speaker 1, given as,

$$\Delta d = d_2 - d_1 \tag{5.7}$$

1. **Constructive interference**: if the loudspeakers are placed one wavelength apart, the crests and troughs of both waves will be aligned. Hence the two waves will be in-phase and will interfere constructively. Therefore, at any time, their path length (Δd) will be a whole number, say n, of wavelengths (λ),

$$\Delta d = n\lambda \tag{5.8}$$

where, $n = 0, 1, 2, 3....$

2. **Destructive interference**: when the speakers are separated by half a wavelength, the incoming waves will be out-of-phase. The sum of the two waves is zero at every point, which is destructive interference. The path length will be,

$$\Delta d = \left(n + \frac{1}{2}\right)\lambda \tag{5.9}$$

where, $n = 0, 1, 2, 3....$

Experiment-to-do:

Explore constructive and destructive interference with water and sound waves using the PhET simulator (figure 5.15).

1. **Step 1: Water waves:**
   (a) Open the PhET simulation and select the **water** tap.

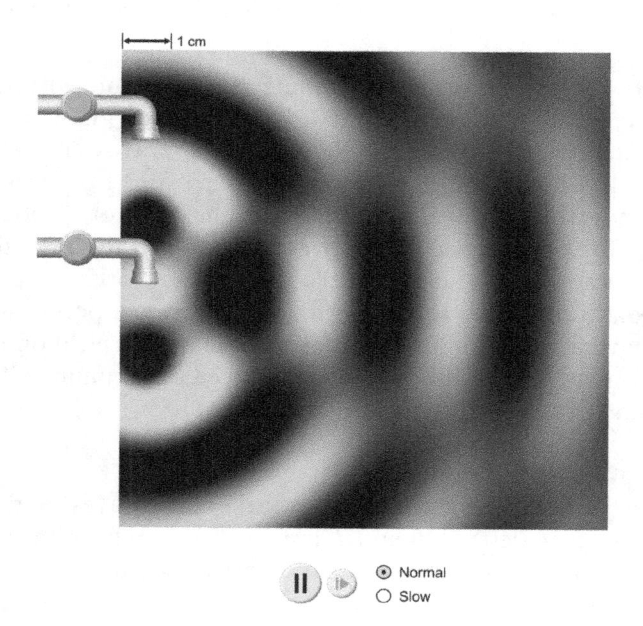

Figure 5.15. Water wave interference pattern.

(b) Find points of constructive and destructive interference on the screen. Which points are constructive? Which are destructive?

(c) Change the distance between resources to 1 cm, 2 cm, 3 cm, 4 cm and 5 cm, and observe changes in the interference patterns. Draw what you observe here or insert screen shots.

(d) Change the amplitude (mid. and max.) of water waves and observe what changes or remains the same.

(e) Next change the frequency and keep the amplitude at max. Observe the changes and explain in your own words (using nodes and anti-nodes) with drawings what you see.

2. **Step 2: Sound waves:**
   (a) For open the simulation, this time select the **sound** tap.

   (b) Create an interference pattern with sound waves. Draw a picture of observations here.

(c) Find points of constructive and destructive interference on the screen. What do you observe?

(d) Change the separation to 100 cm, 200 cm, 300 cm, and 400 cm and observe changes in the patterns. Copy what you observe here or insert screen shots.

(e) Change the amplitude (mid. and max.) of sound waves and observe what changes or remains the same.

(f) Next, change the frequency and keep the amplitude at max. Observe the changes and explain in your own words (using nodes and anti-nodes) with drawings what you see.

5.15 Activity-to-do: beats

Materials needed:

Two tuning forks of close frequencies, within 4 Hz if possible; two people; rubber mallet or another soft surface.

Background:

When two sound waves of different but close enough frequencies approach your ear, the alternating constructive and destructive interferences causes the sound to be alternately soft and loud, a phenomenon called '*beating*' or producing beats. The beat frequency is also called the *rate of the throbbing*. The number of beats per second equals the difference in frequency between the two interfering waves. The slow periodic variations in the amplitude at frequency (Δf) are called *beats*. The beat frequency for the two waves with frequencies f_1 and f_2 is defined as:

$$f_{\text{beat}} = \Delta f = |f_2 - f_1| \tag{5.10}$$

As long as the frequency difference Δf is less than about 10 Hz, the beats are easily perceived. The perceived frequency is the average of the two frequencies, given mathematically as,

$$f_{\text{perceived}} = \frac{(f_1 + f_2)}{2} \tag{5.11}$$

It is worth noting here that $f_{\text{perceived}}$ determines the pitch, whereas f_{beat} determines the frequency of the modulations.

A tuning fork is a metal device that has a handle and two prongs which are called '*tines*'. When you strike a tuning fork against a soft surface, for example the bottom of a shoe, the tines vibrate back and forth depending on the length of the times. The shorter the length the higher the vibrations and hence higher the frequency of the tuning fork.

Example 5.5. Two tuning forks with slightly different frequencies are struck simultaneously and the sound waves travel to our ear. Let the first tuning fork be 512 Hz and the second tuning fork 506 Hz. What will we hear?

Solution:

Experiment-to-do:

1. The two partners in a team.
2. Stand with your back to your partner. Close your eyes.
3. Have your partner strike one tuning fork on the rubber mallet and hold it up to your ear. What do you hear?

4. Next, have the partner strike the other tuning fork on the rubber mallet and hold it up to your ear. What do you hear?

5. Now, have your partner strike both tuning forks on the rubber mallet and hold both simultaneously up to your ear. Describe what you hear.

IOP Publishing

The Physics of Sound and Music, Volume 2
A complete course text (Lab manual)
Samya Bano Zain

Chapter 6

Damping and resonance in musical instruments

6.1 What to know about resonance and damping

1. Simple harmonic motion—mass on a spring
2. Sea in a sea shell
3. Damping
4. Damped harmonic oscillators
5. Classes of damped oscillators: under-damped, over-damped, critically damped
6. Resonance
7. Sympathetic vibrations
8. Helmholtz resonator
9. Chladni patterns.

doi:10.1088/978-0-7503-6350-1ch6

Example 6.1. Suppose a certain spring that is hanging from a hook from the ceiling, stretches 10 cm when it is loaded with a 1 kg mass. Find:

1. The spring constant (k).
   Solution:

2. Period of vibration (T_m).
   Solution:

3. Frequency of vibration (f_m).
   Solution:

Example 6.2. Suppose a 2 kg mass resting on a table is attached to a spring. The coefficient of friction between the mass and table is 0.1 and the spring constant is 20 N m^{-1}. Find the period of an under-damped real pendulum if the oscillations die out after about 3 seconds.

Solution:

6.2 Activity-to-do: 'Seeing' simple harmonic motion—mass attached to a spring

Materials needed:

A mass attached to a spring; horizontal source of light; a vertical surface for light to be projected onto, e.g. vertical wall or blackboard.

Experiment-to-do:

There is an easy way to produce simple harmonic motion by using a mass attached to a spring and some lights (figure 6.1).

1. Attach the mass to a spring and suspend it from a hook or stand.
2. Project light onto to system from the front.
3. Pull the mass down by 5 cm and let it swing.
4. Project the motion of the mass onto the 'projecting surface'.
5. Record with your phone the movements of both the mass-string system and the projection.
6. Repeat the process with increasing displacements in equal increments. What do you observe?

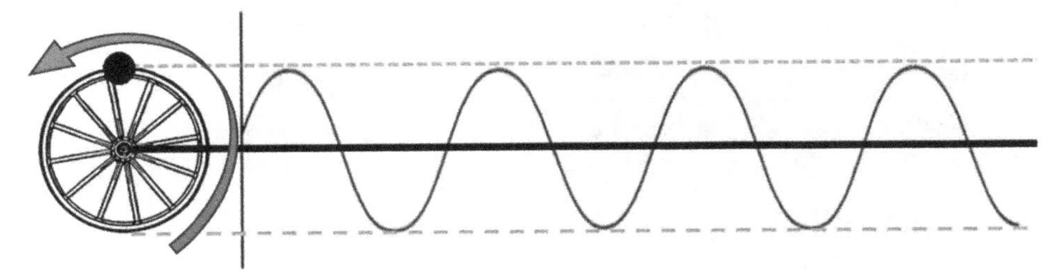

Figure 6.1. Projection of circular motion is simple harmonic motion.

6.3 Activity-to-do: spring force—calculate the period (T)

6.3.1 Find the spring constant (k)

Materials needed:

A spring; measuring ruler; hanger; various masses (50–500 gms).

Background information:

Calculate the spring constant of our spring. According to Newton's third law, for a body in the equilibrium state, the upwards restoring force (F_R) must equal the weight hanging on the spring (mg). As the mass increases, the resulting force also goes up. Mathematically, the magnitude of the restoring force is,

$$F_R = k\Delta y$$

where, k is the spring constant and Δy is the displacement from the equilibrium position. This is called Hooke's law, after Robert Hooke, who stated that 'The power of any springy body is in the same proportion with the extension.' However, the wording has been corrected/updated by replacing the word power with the word force (figure 6.2).

Experiment-to-do:

1. Hang the spring from a hook and use the mass hangers to add masses to the other side of the spring.
2. Start from the smallest mass (m_1) and take five readings in equal increments.
3. Put values in the table 6.1.

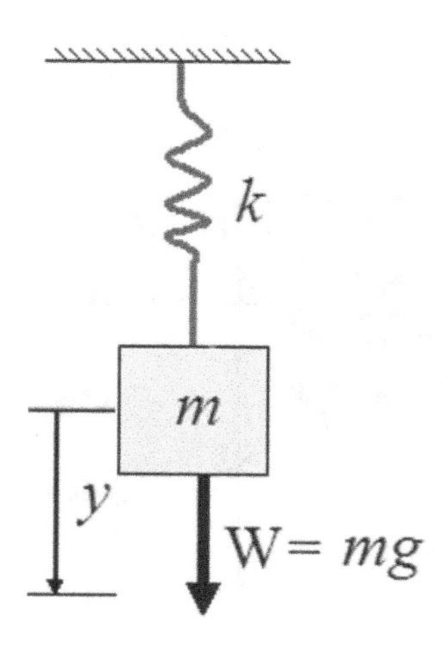

Figure 6.2. Newton's third law, restoring force (F_R) equals weight (mg).

Table 6.1. Find spring constant (k).

	Mass on spring (kg)	Actual position on ruler (cm)	Δy (cm)	Δy (m)	Spring constant (k) (kg s^{-2})
1					
2					
3					
4					
5					
			Average k =		

4. To find spring constant (k), use,

$$k = \frac{mg}{\Delta y}$$

5. What do you observe?

6.3.2 Calculate the period (T_m) of the mass–spring system

Materials needed:
A spring; hanger; stop watch; measuring ruler; various masses.

Background information:
For predicted period of the mass–spring system use the formula below and the value of k obtained in activity 6.3.1,

$$T_m = 2\pi\sqrt{\frac{m}{k}}.$$

Example 6.3. Determine the period of a mass–spring system whose spring constant (k) is 25 Kg s^{-2} and it stretches 2 cm under a certain mass.
Solution:

Experiment-To-Do:

Measure the period (evaluated period) of the mass–spring system.

1. Measure the mass of the spring.
   Mass of spring = ________
2. Hang one side of the spring from a hook and use the mass hangers to add masses to the other side.
3. Start from the smallest mass (m_1) and add it to the hanging spring.
4. Move the mass up by 5 cm and gently let it go. It will start to move up and down under the influence of gravity.
5. Measure total time for 10 complete swings. Hint: To eliminate one source of error, do not use the first couple of swings.
6. The period of the mass–spring system is then the total time measured for 10 complete swings divided by 10.
7. Repeat this three times and average your total times.
8. Record the data in table 6.2.
9. Repeat the above steps by increasing the masses and record the data in table 6.2.

Table 6.2. Calculating period (T_m) of mass–spring system.

	Mass on spring (kg)	Mass of spring (kg)	Total mass (kg)	Predicted period (s)	No. of swings	Total measured time (s)			Average time (t_{av})	Evaluated period (s)
						Time 1 (t_1)	Time 2 (t_2)	Time 3 (t_3)		
1.										
2.										
3.										
4.										
	Average predicted period =					Average evaluated period =				

Discussion questions:

1. What do you observe as you increase the mass on the spring?

2. How does the period (T_m) depend on the mass on the spring?

3. Calculate your % error in evaluation of T_m.

4. How close are you to the actual value of the period? Explain what could be the sources of error in your calculations.

5. What is the SI unit of the period?

6.4 Activity-to-do: damped simple harmonic oscillator (SHO)

6.4.1 Damped SHO—mass–spring system

Materials needed:

Motion detector, mass-and-spring system; logger-pro software; various masses.

Background information:

A real vibrating system has dissipative forces such as friction. As such, the system loses mechanical energy. If there is no continuous driving force it will use up the initial energy and over time its amplitude decreases and eventually the oscillating system comes to rest at its natural equilibrium state. This is called '*damping*' and is shown in figure 6.3.

Damped oscillators can be characterized broadly into three types: under-damped, over-damped and critically damped oscillators. For under-damped oscillators friction is small, for example a pendulum moving through air, whereas friction is quite large in critically damped oscillators and is extremely large under over-damped conditions. An over-damped system released from rest usually moves to the equilibrium point without going past it, whereas in a critically damped system displacement falls quickly and the system may pass the equilibrium point once before gradually relaxing to rest.

The frequency of an under-damped real system (f_{ud}) is,

$$f_{\text{ud}} = \frac{1}{2\pi}\sqrt{4\pi^2 f_R^2 - a^2}, \tag{6.1}$$

where, f_R is the resonance frequency under ideal conditions ($f = 1/T$) and a is the coefficient of damping, calculated by,

$$\text{Damping coefficient } (a) = \frac{\mu}{2m}, \tag{6.2}$$

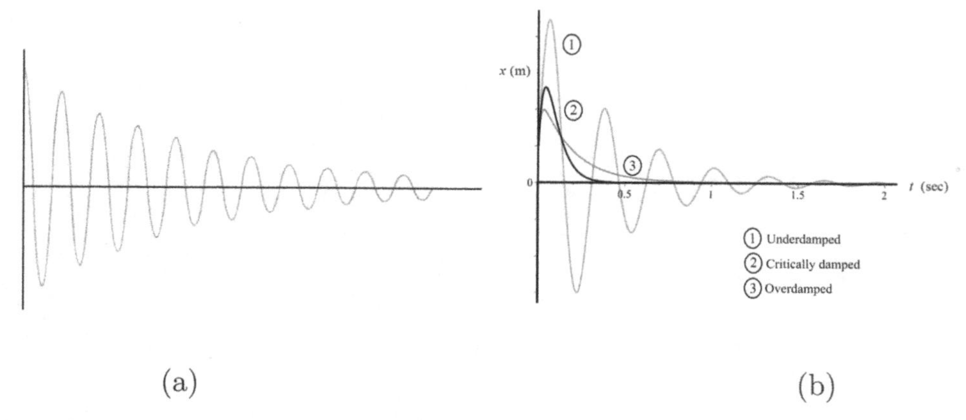

Figure 6.3. (a). Damped simple harmonic motion. (b) Graphical description of the various types of damping.

where, μ is the coefficient of friction and m is the mass of the object under consideration.

Example 6.4. Suppose a 2 kg mass resting on a table is attached to a spring. The coefficient of friction between the mass and table is 0.1 and the spring constant is 20 N m^{-1}. Find the period of an under-damped real pendulum if the oscillations die out after about 3 seconds.

Solution:

Experiment-to-do:

Use a motion detector to record the motion data for a mass attached to a spring.

1. Suspend the mass from the spring.
2. Put the motion detector on the table, facing upwards.
3. Zero the motion detector and change the time axis to 5 minutes and begin data collection by using the green 'collect' button.
4. Put the smallest mass (m_1) on the scale, move it down by 3 cm and allow it to oscillate up and down.
5. What do you observe on the screen? Draw the distance versus time graph here.

6. How long does it take for the spring to stop oscillating?

7. Suspend the next heavier mass (m_2) from the spring. Move it down by 3 cm and allow it to oscillate up and down. What do you observe on the screen? Draw the distance versus time graph here.

8. Suspend the heaviest mass (m_3) from the spring. Move it down by 3 cm and allow it to oscillate up and down. What do you observe on the screen? Draw the distance versus time graph here.

6.4.2 Damped SHO—PhET simulation

Materials needed: PhET simulation (https://phet.colorado.edu/sims/html/waveon-a-string/latest/wave-on-a-string_en.html Wave on a string simulation).

Experiment-to-do: Open the simulation, (wave on a string) (https://phet.colorado.edu/sims/html/waveon-a-string/latest/wave-on-a-string_en.html Wave on a string simulation).

1. Click on 'restart' and change the settings to the following: oscillate, no end, medium tension, slow motion, amplitude (=0.50 cm), frequency (=1.00 Hz).
   (a) **Part 1:** Change the 'damping' setting to 'zero damping'.
   (b) What do you observe on the screen? Draw your observation here.

   (c) Would you call this situation Under-damped/Over-damped/Critically damped? Give reasoning.

2. Click on 'restart' and change the settings:
   (a) **Part 2:** Change the 'damping' setting to 'medium damping'.
   (b) What do you observe on the screen? Draw your observation here.

   (c) Would you call this situation under-damped/over-damped/critically damped? Give reasoning.

3. Click on 'restart' and change the settings.
   (a) **Part 3:** Change the 'damping' setting to 'lots damping'.
   (b) What do you observe on the screen? Draw your observation here.

   (c) Would you call this situation under-damped/over-damped/critically damped? Give reasoning.

6.4.3 Activity-to-do: resonance—sea in a sea shell

Materials needed:
Different sea shells.

Background information:
The word resonance comes from Latin and means to '*resound*', translated as 'to sound out together with a loud sound'. Resonance occurs when the frequency (f) of the forced oscillation matches the natural frequency (f_0) of the system. This results in the maximum transfer of energy to the oscillator, and the amplitude of the system steadily grows until it reaches a maximum A_{max}. The numerical value of A_{max} depends on the damping forces affecting the oscillator (figure 6.4).

Experiment-to-do:
Hold the sea shell to your ear, what do you hear? Please explain in your own words.

Figure 6.4. Seashells.

6.5 Sympathetic vibration

6.5.1 Activity-to-do: example of sympathetic vibration

Materials needed:

Sympathetic tuning forks set; rubber mallet or other soft surface.

Background information:

Sympathetic vibrations are vibrations that are set up in a material when another vibrating material is brought close to it without having direct contact between the two materials. The media between the two materials could be air or any other material. It is important to note that the second media will only have the frequency of the original vibrating material.

A good example of the sympathetic vibration is the soundboard used in musical instruments. Most instruments use vibrating strings that are plucked or struck. When the string vibrates, it forces the soundboard to vibrate at exactly the same frequency. The soundboard transforms the low sound produced by the plucked string into a louder sound since it moves a much greater volume of air. In other words, the resonant properties of the soundboard increase the loudness of the instrument when compared to the loudness of the string alone.

Experiment-to-do:

1. Strike the first tuning fork on the rubber mallet or the sole of your shoe to produce sound (figure 6.5).

2. Lightly touch the wood of the first tuning fork after striking it with the rubber mallet. What do you feel?

Figure 6.5. Sympathetic vibrations in tuning forks.

3. Strike the first tuning fork on the rubber mallet again and bring it close to the other tuning fork, taking care not to touch the two tuning forks. Touch lightly the second tuning fork wood base. What do you observe?

4. Strike the first tuning fork on the rubber mallet again and bring it close to the other tuning fork. Dampen the first tuning fork and hold your ear to the second tuning fork. What do you observe?

5. Explore what happens when two tuning forks sound at different frequencies instead of identical ones?

6.5.2 Activity-to-do: setting-up sympathetic vibrations

Materials needed:

Two or three tuning forks (250 Hz, 512 Hz; 1000 Hz, 2000 Hz etc); ping-pong ball; light string; tape; ruler.

Experiment-to-do:

This activity explores how sound energy (vibrations) are transferred from the tuning fork to the ping-pong ball through air to make it move.

1. Cut a string about 50–60 cm in length.
2. Attach the string to the ping-pong ball using a piece of tape.
3. Hang the string from a hanger.
4. Without striking the tuning fork, move it slowly toward the ping-pong ball without touching the ping-pong. What do you observe?

5. Next, strike the first tuning fork to produce sound.
6. Then slowly bring it close to the ping-pong ball, taking care not to touch the tuning forks and ping-pong ball. What do you observe? Hint: it might help to record this process with a slow-motion camera.

7. Repeat the same process with the other tuning forks and record your findings in table 6.3.
8. What do you observe? Is there a relationship between the frequency of tuning fork and the distance the ping-pong ball moved?

Table 6.3. Setting-up sympathetic vibration.

	Frequency of tuning fork (Hz)	Distance the ping-pong ball moved (cm)
1.		
2.		
3.		
4.		

6.6 Activity-to-do: Helmholtz resonator

6.6.1 Activity-to-do: make your own Helmholtz resonator

Materials needed:

Three empty plastic or glass narrow-neck bottles; ruler; water, marker.

Background information:

The Helmholtz resonator, as shown in figure 6.6(a), is a device created in the 1850s by a German physicist Hermann von Helmholtz (1821–94), and it is used to identify various frequencies present in music and other complex sounds. Helmholtz resonance is the phenomenon of air resonance in a cavity, such as when one blows across the top of an empty bottle. The resonant frequency is determined by the size and the shape of the resonator. This activity uses water bottles to investigate how the volume of air in the bottle affects the pitch of the note that it makes. Please note that a Helmholtz resonator is not the same as an open-ended air column, where the entire container has the same width and the resonant frequency only depends on the container's length.

Experiment-to-do:

1. Blow across the tops of the bottles by touching your lower lip to the edge of the bottle, pursing your upper lip, and blowing gently over the opening. How does the bottle sound?

2. Leave one bottle empty, call it bottle 1.
3. Fill bottle no. 2 up to exactly the half-way mark with water, call it bottle 2.
4. Fill bottle no. 3 to exactly three-quarters up with water, call it bottle 3.

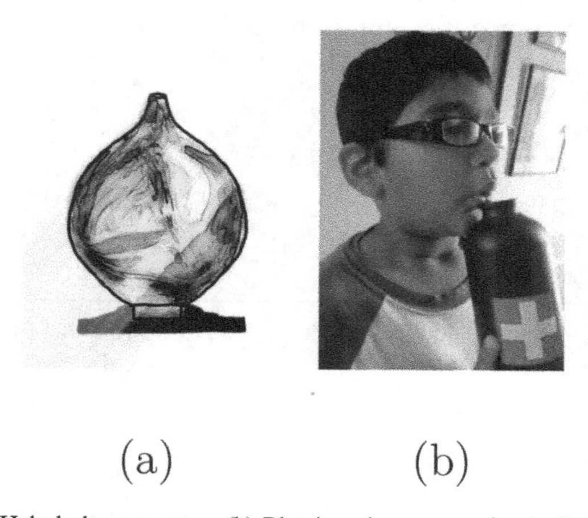

(a) (b)

Figure 6.6. (a) A Helmholtz resonator. (b) Blowing air across a simple Helmholtz resonator.

5. Blow across all three bottles. Keep trying until you hear a clear note. Hint: It may take some practice to make a note from bottle 3.
6. How does the note made by bottle 1 compared to bottle 2?

7. Which note is higher in pitch?

8. How does the note made by bottle 2 compared to bottle 3? Which has higher pitch?

9. How do the notes made from the three bottles compare to each other?

10. Explain in your own words what you observed happened to the frequency of vibration and therefore pitch as you added water to the bottle.

Alternative/additional activities: Further exploration activity ideas:

1. If you have access to other musical instruments, like piano or a keyboard, compare the notes set up in the water bottles to the notes on a real instrument.
2. Does the material of the bottle, shape or size of the bottle affect the note it makes? Redo this activity with glass/plastic bottles of different shapes and sizes. What do you observe?
3. If you have access to narrow-necked glass bottles tapping the bottles below the waterline with a wooden mallet to produce the sound. What do you hear? Is there a difference in the note produced by tapping the bottle instead of blowing over the top of the bottle?

6.6.2 Activity-to-do: frequency of a Helmholtz resonator

Materials needed:

A set of Helmholtz resonators or plastic/glass bottles; calculator; measuring devices, (ruler, vernier caliper etc); an application to detect the frequency of a sound (example, online; for iPhones; for Androids).

Background information:

Helmholtz used a large number of Helmholtz resonators in the 19th century to verify the existence of harmonics in complex tones. By holding successively smaller resonators, with higher and higher frequencies, to his ears as a musical note played into a large opening, Helmholtz was able to hear an increase in amplitude of any frequency present in the harmonic structure of the instrument. From this he could roughly determine the spectrum of the note being played. The Helmholtz resonator is very important because it possesses a single low isolated resonant frequency as opposed to a stretched wire, an open or closed tube or other complex vibrating system. When a Helmholtz resonator is excited by sounds, the resonator causes an increase in amplitude of the very narrow band of frequencies near its resonant frequency and that is what you hear.

Frequency of the Helmholtz resonator:

Suppose the air in the neck of a Helmholtz resonator has an effective length (L) and cross-sectional area (A), the frequency of the Helmholtz resonator will be mathematically given as,

$$f_{\mathrm{H}} = \frac{v}{2\pi}\sqrt{\frac{A}{VL}} \tag{6.3}$$

where v is the speed of sound in air and V is volume of a spherical cavity.

Example 6.5. Suppose we have a 1 liter bottle with an effective length of neck (L) of 5 cm and cross-sectional area (A) of 3 cm^3. Find the resonant frequency.

Solution:

Experiment-to-do:

In this activity, we will find the frequency of a Helmholtz resonator both mathematically (using equation (6.3) and experimentally by following these instructions:

1. Find the speed of sound in air depending on the temperature of the room.

$$v = 331.4 + 0.6T_{\mathrm{C}}$$

where T_{C} is the temperature of the room in Celsius (°C = $\frac{9}{5}$(°F − 32)).

Speed of sound = _______

2. Find the cross-sectional area of the neck, in m^2.
   Cross-sectional area of the neck = _______
3. Find the volume of the cavity, in m^3.
   Volume of the cavity = _______
4. Find the effective length (L) of the neck. The depth of the opening may take into account the end correction of a hole in acoustics: that's a factor added to the length to justify the different starting point of the sound wave in the opening itself.

$$L = L_0 + \Delta L \approx L_0 + 1.5r$$

   Where L_0 is the true length of the opening and ΔL is the end correction and r is the circular radius of the neck. Calculate the circular radius of the neck by finding the inner diameter of the neck using a vernier caliper and dividing it by 2.
   Effective length (L) of the neck = _______
5. Knowing that the geometry of the resonator and the speed of sound is assumed constant, then the frequency will vary with the cavity volume as,

$$f_H = C_1\sqrt{\frac{1}{V}} \tag{6.4}$$

   where C_1 is a constant that included the speed of sound and the neck geometry.
6. Find C_1 using the formula,

$$C_1 = \frac{v}{2\pi}\sqrt{\frac{A}{L}}$$

   Constant (C_1) = _________
7. To find volume (V), use the following formulas that depend on the shape of your Helmholtz resonator:
   (a) For cylindrical bottles:

$$V = \pi(R^2 - r^2)(H)$$

   where, R is the external radius, r is internal radius of the bottle and H is the height of bottle. To find R, use a string to measure circumference (c) and then use $r = \frac{c}{2\pi}$. In most cases, $R^2 - r^2 \approx R^2$.
   (b) For spherical bottles:

$$V = \frac{4}{3}\pi R^3$$

   where, R is the radius of the bottle. To find R, use a string to measure circumference (c) and then use $r = \frac{c}{2\pi}$.

Volume (V) of Helmholtz resonator = _________

8. Use equation (6.4) to calculate the frequency the Helmholtz resonator.
   Calculated frequency the Helmholtz resonator (f_H) = ...

Table 6.4. Frequency the Helmholtz resonator (f_H).

	f_H (calculated)	f_H (measured)	% difference
1. Empty bottle			
2. Half-way filled			
3. Two-thirds filled			

9. Next, use the online application (online; for iPhones; for Androids) to detect the frequency of sound, record in table 6.4. **Measured frequency the Helmholtz resonator** (f_H) = ______.
10. Fill the bottle half way with water and repeat the experiment. What do you observe?

11. Next, fill the bottle two-third way up and again repeat the experiment. What do you observe?

6.7 Activity-to-do: singing rods

Materials needed: Singing rod, preferably aluminum (https://www.teachersource.com/product/singing-rod) of at least two lengths (available lengths 60, 80, and 100 cm).

Background information:

Any rod can be made to vibrate along its length at its characteristic frequency thus making it '*sing*'. Aluminum has low density for a metal which makes it a good conductor of sound. Hence the most commonly used rod for this purpose is an aluminum rod.

Experiment-to-do:

1. To make the rod sing, firmly hold the rod at its mid-point with thumb and index finger of say the right hand, called the '*nodal-hand*', as shown in figure 6.7.
2. To find the exact middle, or the center-of-mass, start with your hands at both ends of the rod. Now slowly bring your arms together while allowing your fingers to slide beneath the rod. Since the rod is uniform in shape, your fingers will automatically come together at the center-of-mass of the rod.
3. Next, scrape along the length of the rod by the thumb and pointer finger of the left hand, called the '*scraping-hand*'. It helps if the scraping-hand is dusted (best way is to pinch and release a small amount with the thumb and forefinger) with crushed violin or cello rosin.
4. Gently stroke the rod from the center to the end of the rod using your rosin coated scraping-hand. Repeat while slightly increasing the pressure of your scraping-hand until you hear a high-pitched tone. Sound produced will depend on the pressure applied, too little pressure will not set up vibrations

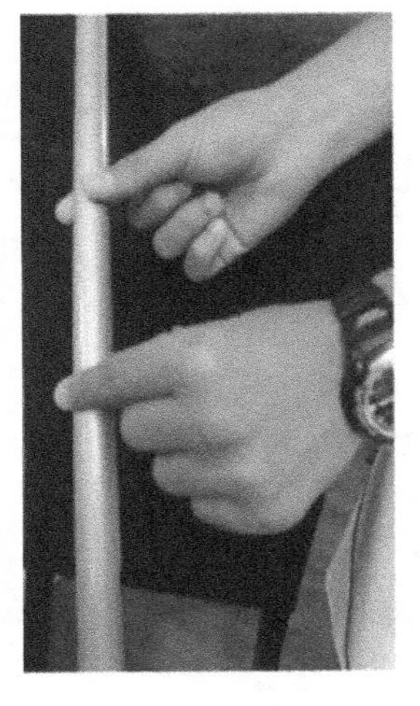

Figure 6.7. Playing a singing rod.

while too much pressure will dampen the sound. Remember practice makes progress!

5. What do you observe?

6. Explain your observation in terms of the pulling motion and the response of the aluminum metal rod.

7. Explain what is happening in figure 6.8.

8. Redo this experiment by rubbing the rod more and more forcefully. What do you observe? Is the sound thus produced pleasant or unpleasant?

9. Redo this experiment with longer rods. What do you notice about the sound that is produced? Is it higher pitched or lower pitched than the first shorter rod?

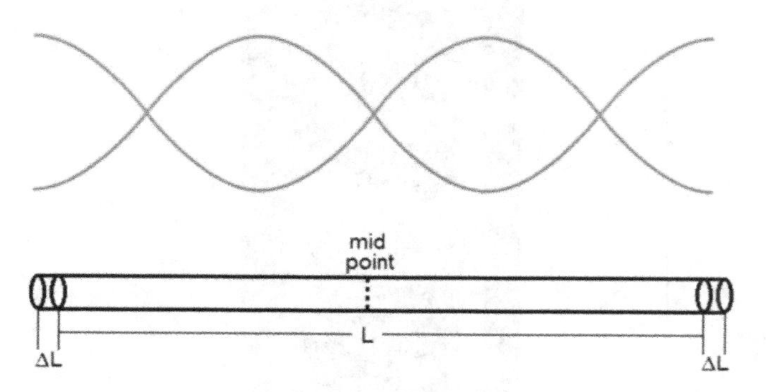

Figure 6.8. Schematic of a singing rod, with nodes and anti-nodes.

Part III

Sound propagation

IOP Publishing

The Physics of Sound and Music, Volume 2
A complete course text (Lab manual)
Samya Bano Zain

Chapter 7

Sound propagation

7.1 What to know about sound propagation

1. Waves—traveling versus standing waves (again!)
2. Properties of traveling waves; amplitude, wavelength, period, frequency, speed of transverse waves
3. Waves on a string: speed, kinetic energy, elastic potential energy, power
4. Speed of sound waves
5. Factors influencing speed of sound; temperature, medium
6. Speed of sound waves through solids, liquids and gas.

doi:10.1088/978-0-7503-6350-1ch7

7.1.1 Activity-to-do: transmission of a wave

Materials needed:

A round bowl about half full of water, ground black pepper, dishwashing soap.

Background information:

A wave is characterized as a disturbance that travels away from its source. The particles of the medium a wave travels through are disturbed from their equilibrium position as the wave passes and then return to their equilibrium position after the wave has passed. Traveling waves, also called '*progressive waves*' are waves that travel. They occur when a wave is not confined to a given space along the medium. The most commonly observed traveling wave is an ocean wave.

Experiment-to-do:

This activity shows how waves emanate outwards from where the initial disturbance occurred.

1. Sprinkle the pepper into the bowl of water until there is an even amount floating on the surface.
2. Carefully allow one drop of soap to fall in the middle of the bowl.
3. What do you observe?

4. Explain in your own words that is happening?

7.1.2 Activity-to-do: wave on a string

Materials needed:

PhET simulation (https://phet.colorado.edu/sims/html/wave-on-a-string/latest/wave-on-a-string_en.html Wave on a string simulation).

Background information:

Suppose a sinusoidal wave on a string travels in the positive x-direction, as set up in figure 7.1. The wave moves through the string, elements of the string oscillate up and down parallel to the y-axis. At time (t), the displacement along the y-axis is called Ψ and is mathematical represented by,

$$\Psi(x, t) = \mathrm{A}\sin(kx + \omega t) \tag{7.1}$$

where, A is the amplitude of the wave, x is the position of a point along the string, t is the time and ω is the angular frequency ($\omega = 2\pi f$, where, f is frequency).

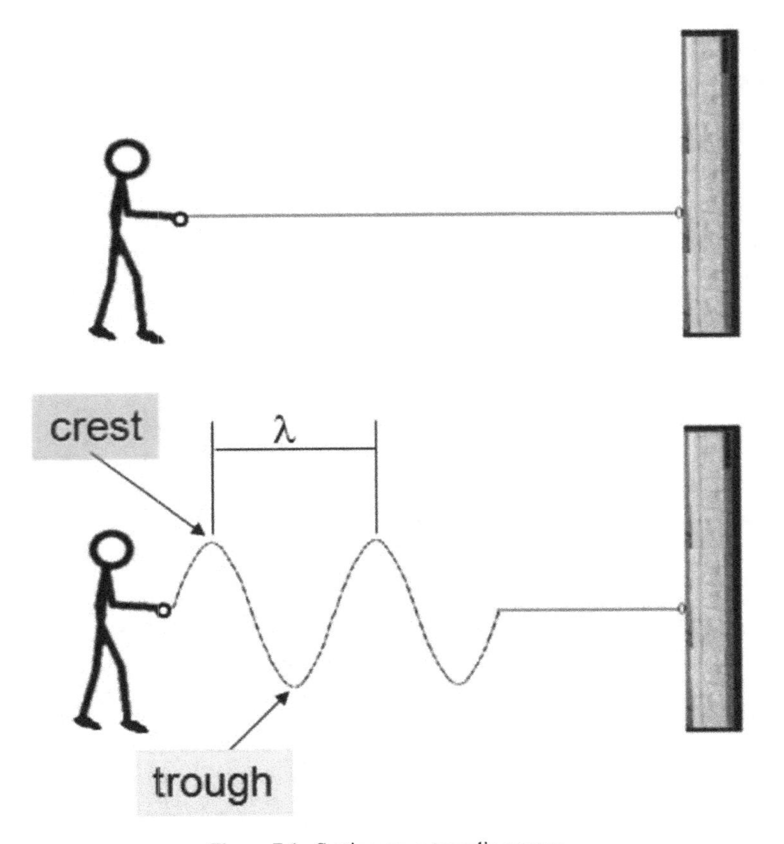

Figure 7.1. Setting up a traveling wave.

Example 7.1. A transverse wave traveling along a string the x-axis has the form:

$$\Psi(x, t) = y(x, t) = (3\ \text{m})\sin(72(\text{rad m}^{-1})x + 2(\text{rad s}^{-1})t)$$

What is the frequency, angular frequency, wave number, period, amplitude of this wave?

Solution:

Speed of transverse waves on a string:

The speed of a wave depends on the properties of the medium. For a string of length (L) and mass (m), held under a tension (T), the speed of the wave is given as,

$$v = \sqrt{\frac{TL}{m}} = \sqrt{\frac{T}{\mu_l}} \tag{7.2}$$

where, μ_l, is a quantity called *linear mass density*. This equation shows that the speed of the wave is independent of the frequency of the wave.

Example 7.2. A transverse sinusoidal wave (with $f = 100$ Hz) is traveling along a string with linear density $\mu_l = 0.5$ kg m^{-1} that is held under tension $T = 50$ N. If the maximum amplitude $A = 1$ cm, find the average rate of energy transport.

Solution:

Experiment-to-do:

Open the PhET simulation (https://phet.colorado.edu/sims/html/waveon-a-string/latest/wave-on-a-string_en.html Wave on a string simulation), and perform the following experimentation (figure 7.2).

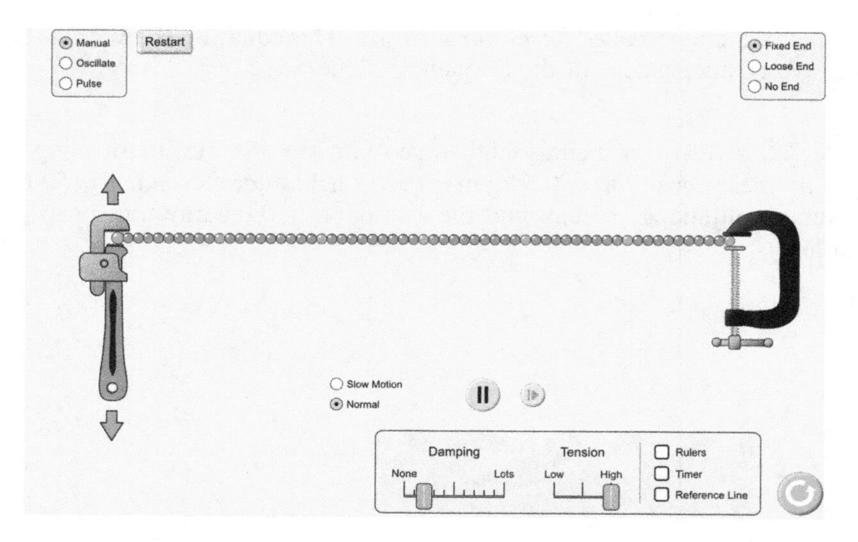

Figure 7.2. PhET simulation—wave on a string.

- **Part 1: Introduction and discovery:**
  1. On the 'manual' setting, move the wrench up and down. Explain in your own words what you observe in each case.

  2. Click 'Oscillate' on the upper left corner. What do you observe?

  3. Change to 'Pulse mode' on the upper left corner and keep the 'amplitudes' at medium and vary the 'Pulse width' option to minimum and maximum values. Draw and explain in your own words what you observe in each case.

4. Try both options at 'Slow Motion' and 'Normal' option. What do you observe?

5. Next, change both the 'amplitudes' and 'Pulse width' options to minimum and maximum values. Explain in your own words what you observe in each case.

- **Part 2: Evaluating frequency:** Click on 'Restart' and change the settings to the following: Oscillate, amplitude (= 0.75 cm), frequency (= 1.50 Hz), zero damping, high tension, Slow Motion, Timer.
  1. Part 2a: Select 'Fixed End': hit play. What do you observe?

  2. Part 2b: Select 'Loose End': pause the simulation and change the setting to 'Loose End'. Hit play and let the simulation run for at least 10 seconds. What do you observe?

  3. Part 2c: Select 'No End': pause the simulation and change the setting to 'No End'. Hit play and let the simulation run for at least 10 seconds. What do you observe?

- **Part 3: Relation between frequency and number of waves.** Pause and restart the simulation. Reselect 'Fixed End' and use the timer function, count how many crests (middle green ball at maximum point) pass the center line in 15 seconds. Record in table 7.1,

Table 7.1. Relation between frequency and number of waves.

Frequency (Hz)	Number of crests in 15 s	Number of waves in 1 minute
0		
0.25		
0.75		
1.00		
1.25		

- What do you observe in table 7.1? Given the data you obtained, can you define 'frequency'?

- **Part 4: Measuring amplitude:**
    1. Click on 'Restart' and change the settings to the following:
    2. Oscillate, No End, medium damping, high tension, Slow Motion, Timer.
    3. Turn on the 'Ruler' option and drag the vertical ruler such that the 0 cm mark aligns with the central horizontal line. The location of the middle green should coincide with the vertical ruler.
    4. Move the horizontal ruler out of the way.
    5. Turn the simulation 'on' again. What do you observe?

    6. What is the amplitude of the wave?

    7. Change the 'amplitude' and measure the height of the wave. Hint: make sure to 'restart' the simulation after each data point (table 7.2).

Table 7.2. Relation between amplitude, height and number of waves.

Amplitude (cm)	Height of wave	Number of waves in 15 s
0		
0.25		
0.75		
1.00		
1.25		

8. As you increased the amplitude, what did you observe about the number of waves in 14 s?

9. As you increased the amplitude, what did you observe about the wavelength of the wave?

10. As you increased the amplitude, what did you observe about the frequency of the wave?

11. Explain in your own words that is happening?

7.1.3 Activity-to-do: simulation—dependance of speed of sound on temperature

Materials needed:
Calculator and pen.

Background information:
The speed of sound changes with the temperature of the atmosphere. For instance, speed of sound in air (v_o) at about sea-level at 0 °C is 331 m s^{-1}, whereas at room temperature (T = 20 °C) the speed of sound is about 344 m s^{-1}. There are two equivalent explanations for this phenomenon:

1. Molecules at higher temperatures have more energy, thus they vibrate faster. Since the molecules vibrate faster, sound waves travel through them more quickly.
2. Provided the pressure remains constant, as the temperature increases the density of air decreases. Now as the density decreases air molecules move closer to each other, which results in the increase in the speed of sound.

Experiment-to-do: This activity shows how the speed of sound varies with temperature. Here we will find the speed of sound that corresponds to temperatures recorded around the world.

1. Calculate the speed of sound at temperatures (T) using the equation,

$$v_{\text{sound}} = (331.3 + 0.6T)$$

2. Convert temperature from Fahrenheit (°F) to Celsius (°C),

$$T_{(°F)} = \frac{9}{5}T_{(°C)} + 32$$

3. Record your values in the table 7.3.
4. Plot the speed of sound obtained versus the temperature in °C. Hint: plot temperature on x-axis and speed of sound on y-axis.
5. What does your graph look like? Did you obtain a straight line?

6. Add a trend line (linear) to your graph. Make sure to include the equation by using the 'display equation on chart' option. Write your equation here. What do you observe?

7. Either draw your graph here by hand or include it as a printed copy.

Table 7.3. Relation between temperature and speed of sound.

	Temperature (°F)	Temperature (°C)	Speed of sound (m s^{-1})
1	−40		
2	−30		
3	−20		
4	−10		
5	0		
6	10		
7	20		
8	30		
9	40		
10	50		
11	60		
12	70		
13	80		
14	90		
15	100		
16	110		
17	120		

8. As you increased the temperature, what did you observe about the speed of sound?

9. Explain in your own words why you think it gets quieter as the temperature drops?

7.1.4 Activity-to-do: simulation—dependence of speed of sound on materials

Materials needed:

A bell jar laboratory equipment.

Experiment-to-do:

The speed of sound depends on the medium it is traveling through. It is primarily determined by a combination of the medium's rigidity (or compressibility in gases) and its density. Very dense media have molecules and atoms that are very closely packed together, hence sound will move very fast through them.

1. A bell jar, as seen in figure 7.3 is equipment used to show the dependence of speed of sound on the medium it travels through.
2. Place a bell jar on a base connected via a hose to a vacuum pump.
3. Place an electrical bell in the bell jar.
4. Change the air pressure inside the jar by pumping the air out of the bell jar.
5. Explain how sound is produced by a bell.

6. What is a vacuum?

7. What do you observe as the air is pumped out of the sealed bell jar?

Figure 7.3. A bell in a jar setup, with a vacuum pump assembly.

8. What is the particular vacuum point when no more sound is heard from the bell?.

9. Explain in your words what is happening? What does sound need to propagate?

7.1.5 Activity-to-do: calculate the speed of sound

Materials needed:
Water; resonance tube, tuning forks; thermometer; rubber mallet; calculator.

Experiment-to-do:

1. **Part 1:** Record the temperature of the room using a thermometer. This will be temperature 1 (T_1) for this experiment.
2. Next calculate the speed of sound at temperature T_1 using the equation,

$$v_{\text{sound}} = (331.3 + 0.6T_1)$$

3. **Part 2:** Note that the apparatus is set up properly and the water reaches the maximum height in the resonance tube (if the water does not reach the max value, then put your hand in the water-containing cylinder such that it does). This is your initial position.
4. Strike the highest frequency tuning fork on the rubber mallet and hold it in a horizontal position just above the mouth of the tube.
5. Slide the water cylinder slowly up/down until the note heard from the tube is at its loudest. This is the first resonance point, as shown in figure 7.4(a).

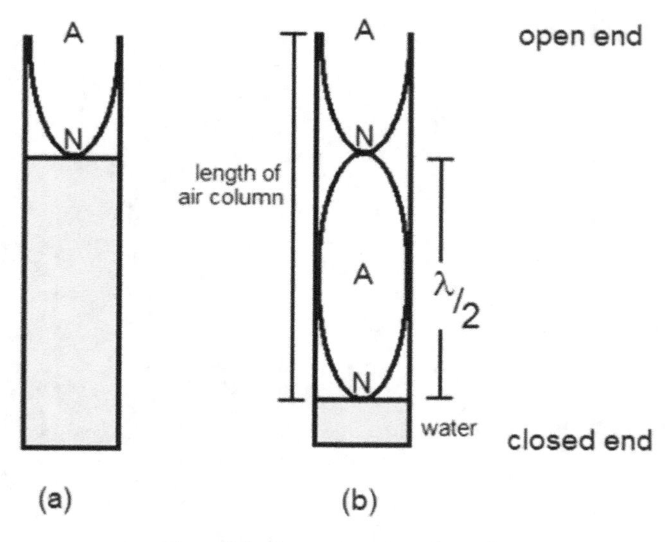

Figure 7.4. Resonance tube schematic.

Table 7.4. Speed of sound.

	Tuning fork frequency Hz	Resonance pt 1				Resonance pt 2				$\lambda = 2(l_{2av} - l_{1av})$ m	v_{sound} m s^{-1}
		$l_{1(1)}$ m	$l_{1(2)}$ m	$l_{1(3)}$ m	$l_{1(av)}$ m	$l_{2(1)}$ m	$l_{2(2)}$ m	$l_{2(3)}$ m	$l_{2(av)}$ m		
1											
2											
3											

6. Measure the length of the air column (from the top of the tube to the water level at resonance point) and put this value in table 7.4 as (l_1).
7. Move the water cylinder such that the air column is 2 or 3 cm less than $3 \times l_1$ (figure 7.4(b)).
8. Find a second resonance point (l_2) with the same tuning fork by again sliding the cylinder up and down until the note heard is at its loudest.
9. Next use your calculator to find the averages for resonant point 1 (l_{1av})and resonant point 2 (l_{2av}).
10. Calculate the wavelength using the formula, $\lambda = 2(l_{2av} - l_{1av})$.
11. Calculate the speed of sound from the formula, $\lambda \times f = v_{sound}$.
12. Record the measurements in table 7.4.
13. Repeat the same steps above to obtain the corresponding values of l_{1av} and l_{2av} for all the tuning forks in order of decreasing frequency.
14. Find the average value for the speed of sound in table. What value do you obtain?

15. Calculate your % error,

$$\% \text{ error} = \frac{\text{Evaluated speed of sound} - \text{Actual speed of sound}}{\text{Actual speed of sound}} \times 100$$

16. How close are you to the actual value of the speed of sound? Explain what could be the sources of error in you calculations?

17. **Part 3:** Repeat the experiment on the coldest day in the winter, when the outside temperature is close to 0 °C and then in the summer when the temperature is as high as possible.
18. Compare the speed of sounds obtained at all three temperatures. What do you observe?

IOP Publishing

The Physics of Sound and Music, Volume 2
A complete course text (Lab manual)
Samya Bano Zain

Chapter 8

Factors impacting sound propagation

In this chapter we will discuss some principles that govern wave behavior, starting with the phenomena that might be considered fundamental to how waves propagate.

8.1 What to know about factors impacting sound propagation

1. Refraction of waves
2. Huygen's principle
3. Diffraction; acoustic diffraction
4. Doppler effect.

8.2 Refraction

Background information:

When a ray of light enters a glass block at an angle other than the normal it changes speed, wavelength and direction. In going from a less dense medium (air) to a denser medium (glass) light bends towards the normal. This means that the angle of incidence is greater than the angle of refraction). In going from a more dense to a less dense medium (glass to air), light bends away from the normal. How much the light bends depends on its color.

The change in angle of the light ray is the same when it enters and leaves the glass. If the incident ray had continued without changing direction, then the emergent ray would be parallel to it. Snell's law says that the ratios of the sines of the angles of incidence and of refraction are constant and this ratio depends on the media under consideration (figure 8.1).

$$n_1 \sin \theta_1 = n_2 \sin \theta_2 \tag{8.1}$$

doi:10.1088/978-0-7503-6350-1ch8

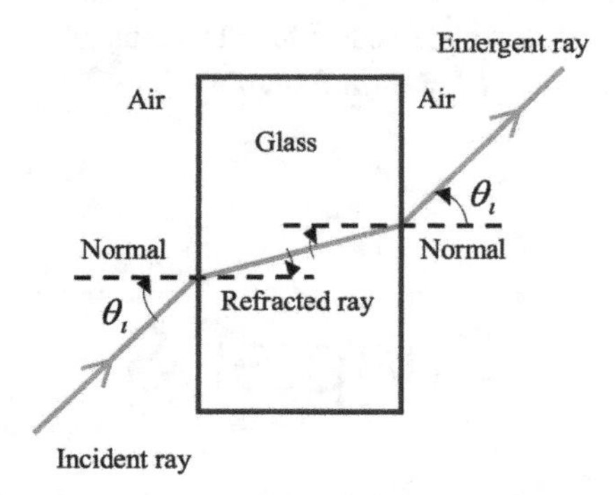

Figure 8.1. Snell's Law, incident ray going through glass, with angle θ_i. Refracted ray inside glass θ_r and emergent ray, with angle equal to (θ_i) is the ray that emits from the glass back into air.

Example 8.1. A light ray travels from air (with $n_\text{a} = 1.0$) into water (with $n_\text{w} = 1.33$). The angle of incidence is 30°. What is the angle of refraction?

Solution:

8.2.1 Activity-to-do: experimental (1): refraction

Materials needed:

Pencil; glass of water; protractor (figure 8.2).

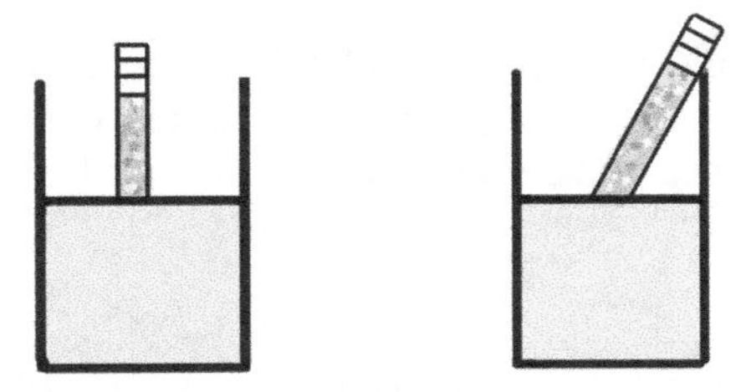

Figure 8.2. Refraction using pencil and water. (Left) pencil held upright. (Right) pencil resting on one edge.

Experiment-to-do:

1. Take a clear glass of water. Hold a pencil upright in it, as seen in figure 8.2 (left). This is your normal line.
2. Observe the pencil from the front of the glass. (Bend down so the glass is at your eye level.)
3. What do you notice about the half of the pencil that is in the water versus the half of the pencil that is outside in the air? Draw your observation on figure 8.2 (left).

4. Next, move the pencil so that it rests against the side of the glass, as seen in figure 8.2 (right).
5. What do you notice about the pencil? Does it look broken when viewed at eye level? What about if you look down the glass from the top?

6. Draw your observation on figure 8.2 (right).

7. Make sure to draw the normal line on the picture and label the angles of incidence and refraction. You do not have to measure the angles from your picture.

8. Take a picture of both set-ups and print them on a printer.
9. On the printed paper draw the normal line on the picture and label the angles of incidence and refraction. You do not have to measure the angles from your picture.

10. If you do not have access to a printer, draw your observations carefully on figure 8.2.

11. Next, use a protractor to measure the angles of incidence and refraction and inert the values here.
    Angle of incidence (θ_i) = ____________.
    Angle of refraction (θ_r) = ____________.
12. What do you notice about the angle of incidence and the angle of refraction?

8.2.2 Activity-to-do: experimental (2): refraction—using pins and laser pointer

Materials needed:
A block of glass, paper, pins, laser pointer, pencil, protractor (figure 8.3).

Experiment-to-do:

1. Put a sheet of white paper on the soft base.
2. Place a plane glass provided to you on the paper and trace the glass and call it line MM′.
3. Mark a point O at the center of the glass surface and draw a normal to the glass surface.
4. Draw IO, the incident ray and fix two pins P_1 and P_2 on the incident ray. Make sure angle $I\hat{O}M'$ is less than 90°.
5. Fix pins P_3 and P_4 on the other side of the glass, to represent the refracted image of P and Q, as observed through the glass as P_3 and P_4 such that they hide the pins P_1 and P_2, as shown in figure 8.4.
6. Remove the pins and glass and draw the incident ray, exit ray and the ray while the light is traveling in the glass
7. Measure the angle of incidence and angles of refractions for the two interfaces. First interface (air–glass) and second interface (glass–air)
8. Write Snell's law for each interface and determine two values of index of refraction for glass. Take n for air = 1.
9. Repeat the experiment for different measures of angle of incidence (table 8.1).
10. **Double check:** use a laser pointer and trace over the incident ray, and draw the refracted image of the laser beam as observed through the glass. (*Don't look directly into the laser, please!*)
11. Does the laser trace the same path as the path drawn when using pins?

Figure 8.3. Verify the laws of refraction using a laser.

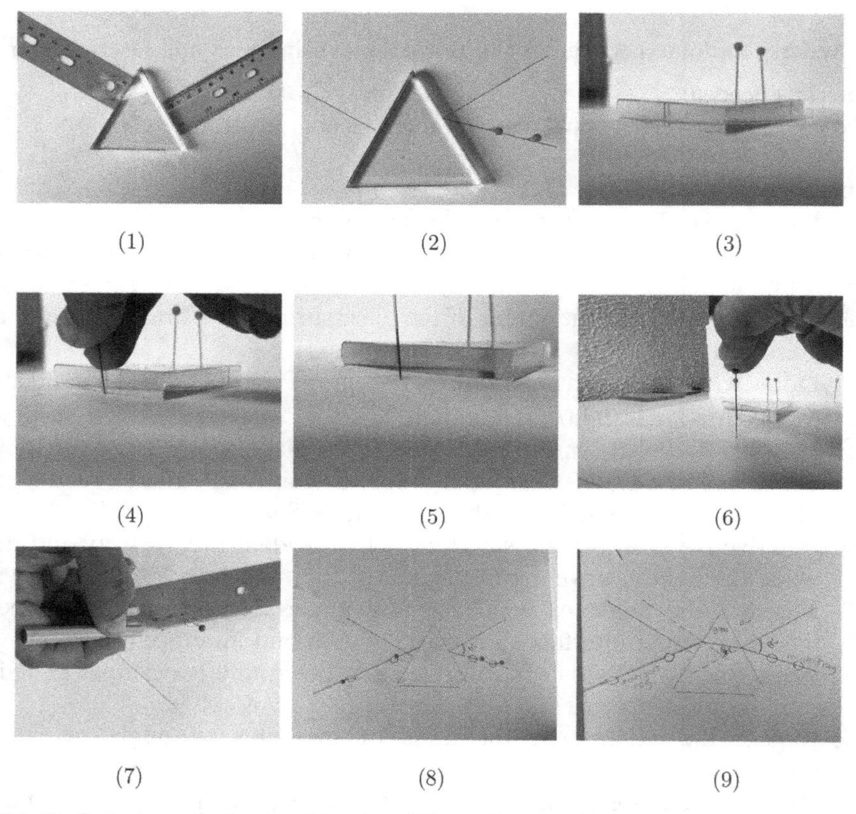

Figure 8.4. Verify the laws of refraction using pins. (1) Draw the normals to the glass using two rulers. (2) Fix two pins (P_1 and P_2). (3) Bend down until you can see pins (P_1 and P_2) through the glass. (4) Line up P_1 and P_2 by moving your head until the two pins become one looking through the glass. (5) Zoomed in view, all pins (P_1, P_2 and P_3) should be in a line. (6) Fix pins P_3 and P_4 such that they cover the pins (P_1 and P_2) while looking through the glass. (7) Draw the emergent ray. (8) Remove pins and glass and draw the refracted and the emergent angles on the paper. (9) Final top view.

Table 8.1. Angles of incidence and angles of refraction.

	First interference (air–glass)			Second interference (glass–air)			
	w.r.t. incident ray			w.r.t. emergent ray			
	Incidence (θ_{1i})	Refraction (θ_{1r})	n_{glass}	Incidence (θ_{2i})	Refraction (θ_{2r})	n_{glass}	% diff
1							
2							
3							
4							

8.2.3 Activity-to-do: experimental (3): refraction—use Snell's law to predict '*n*' for glass using pins

Materials needed:

A block of glass, paper, pins, pencil, protractor.

Experiment-to-do:

1. Put another sheet of white paper on the drawing board.
2. Place a triangle glass on the drawing board and trace the triangle
3. Place two pins on one side of the glass in a line, these two pins (P_1 and P_2) determine the 'incident ray'.
4. Rotate the page so the two pins (P_1 and P_2) are away from you.
5. Place a third pin on the opposite side of the triangle (the side where your eye is) such that it hides the original pins through the glass.
6. Next, hide all three pins with a fourth pin. Pins 3 and 4 determine the 'exit ray' or 'emergent ray'.
7. Remove the glass from the paper and draw the incident ray, exit ray and the ray while the light is traveling in the glass.
8. Measure the angle of incidence and the angle of refraction for the two interfaces. First interface (air–glass) and second interface (glass–air).
9. Write Snell's law for each interface and determine two values of index of refraction for glass. Take n for air = 1.

8.2.4 Activity-to-do: simulation (1): refraction

Materials needed:

Simulator link (https://phet.colorado.edu/en/simulations/bending-light) (PhET, Snell's law); calculator.

Experiment-to-do:

1. Open the simulator link.
2. Click on the icon called 'Intro'.
3. Select substance 1 to be air, and substance 2 to be water.
4. Turn on the laser pointer in the picture, and then drag the protractor into position. Place the center of the protractor exactly at the point where the laser beam hits the interface (figure 8.5).

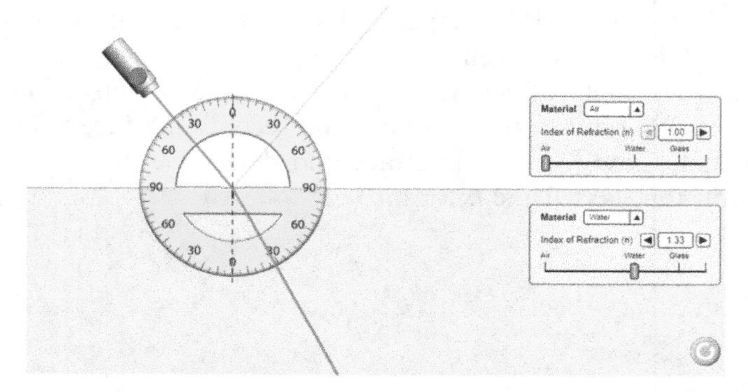

Figure 8.5. Snell's law simulator—Part 1.

5. Adjust the laser beam angle to 40 degrees with respect to the normal and observe the angle of the refracted ray. Does it come out to be the same as the math said it would (yes/no)?

6. Now change the second substance to glass. Keep the first substance air, and θ_1 at 40°. Observe what the simulator gives for θ_2. Use Snell's law to calculate the value for θ_2. Does it match? Show your work and keep a screen shot.

7. Change the setup to make substance one glass and substance two air. Place the laser pointer such that it hits the surface at 40 from the normal. What does the simulator give for θ_2?

8. Calculate θ_2 using Snell's Law to see if it matches what the simulator shows. Be careful to use n_1 as 1.5 and n_2 as 1.0.

8.2.5 Activity-to-do: simulation (2): refraction

Materials needed:

Snell's law simulator (https://phet.colorado.edu/en/simulations/bending-light); calculator (figure 8.6).

Experiment-to-do:

1. Go back to the screen where you clicked on the 'Intro' icon but this time click on the 'prisms' icon at the bottom of the screen.
2. Select the prism option from the various shapes shown in figure 8.6 and position it in front of the laser pointer.
3. Place the prism in a position similar to the figure. The angle doesn't have to be the same as in the picture, however, we do want the beam to come out the other side, not reflecting internally off of the second interface.
4. Use the protractor to see what you get for θ_1 and θ_2 at the first interface? Remember to measure the angle between the light beam and the normal.
   $\theta_1 =$ ____________
   $\theta_2 =$ ____________
5. Use Snell's Law to calculate n_2 (n for glass) (use $n_1 = 1$ for air). Show your work.

6. Now move the protractor to the other side of the prism where the light is coming out. Measure θ_1 and θ_2 at the second interface. Notice that this time light is going from glass to air, so substance one is glass and substance two is air. Your setup should look something like that shown in figure 8.7. Try not to move the prism when you move the protractor to the other side. If it moves, try to put it back where it was.
7. Use Snell's law to calculate n_1 to see what you get for glass (use $n_2 = 1$ for air). Show your equation work and a screen shot here.

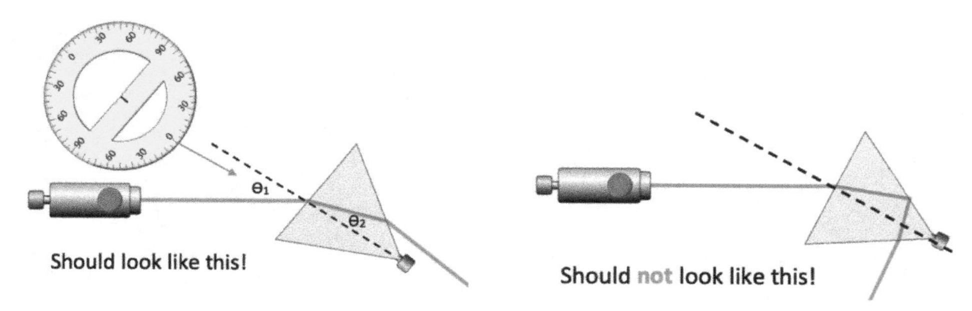

Figure 8.6. Snell's law simulator—Part 2.

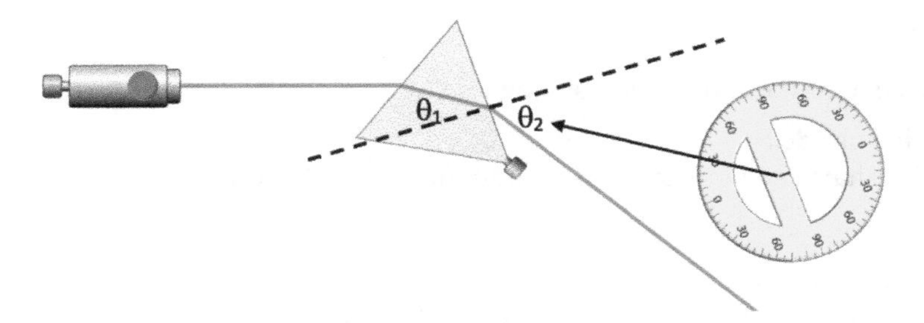

Figure 8.7. Snell's law simulator—Part 2-2.

8.3 Diffraction

8.3.1 Activity-to-do: diffraction grating

Materials needed:

Candle, sharpened pencils, laser, human hair, diffraction grating, prism, color pencils.

Experiment-to-do:

Light bends when it passes around an edge or through a slit. This bending is called diffraction. The diffraction pattern shows that light travels as a wave.

1. **Activity 1: Light the candle—use hair**
   (a) Stretch a hair tight and hold it about 1 inch from your right eye.
   (b) Move the hair until it is between your eye and the candle.
   (c) Note how the light is spread into a line of blobs by the hair.
   (d) Rotate the hair and watch the line of blobs rotate.
   (e) Explain what you observe here.
2. **Activity 2: Light the candle—use pencils**
   (a) Wrap a tape around one pencil to hold the two pencils apart.
   (b) Hold up the two pencils, vertically side by side, such that they form a thin slit just below the tape.
   (c) Hold the pencils 1 inch away from one eye.
   (d) Look at the candle through the slit between the pencils. Notice that there is a line of light perpendicular to the slit.
   (e) Look closely, do you observe that the line of light is actually made up of tiny blobs of light? Draw what you see.

   (f) Rotate the pencils until both are horizontal. What do you observe now?

   (g) Next squeeze the pencils together, making the slit smaller. What do you observe?

   (h) Did the blobs of light become larger or smaller?

(i) Did they become closer to each other or did they grow further apart from each other?

(j) What do you notice about the edges of the blobs? Are they of different colors?

3. **Activity 3: Use a diffraction grating**
   (a) Hold the prism up to different light sources, example, sunlight, yellow light, white light etc.
   (b) What do you see in each case?

   (c) Copy what you observe on a sheet of paper here.

4. **Activity 4: light diffraction through a prism**
   (a) Hold the prism up to sunlight.
   (b) Observe the way sunlight diffracts and separates through a prism.
   (c) How many colors do you see?

   (d) Use color pencils and trace out what you observe.

8.3.2 Activity-to-do: diffraction of sound

Materials needed:
Two team members.

Experiment-to-do:

1. Find a quieter spot in the room, or go outside if it is not too noisy.
2. **Step 1:** Stand facing each other at least 4 to 5 feet part.
   (a) Are you able to see your team member?
   (b) Say 'Good morning' to each other.
   (c) Note how loud the sound appears to you.
3. **Step 2:** Each team member should turn 90° to the right and keep looking straight ahead.
   (a) Are you able to see your team member?
   (b) Say 'Good morning' to each other.
   (c) Did this seem louder, quieter or the same as before?
4. **Step 3:** Next, each team member again should turn 90° to the right and keep looking straight ahead. Hint: At this point both team members should be facing completely away from each other.
   (a) Are you able to see your team member?
   (b) Say 'Good morning' again.
   (c) Did this seem louder, quieter or the same as before?
5. **Step 4:** Again, each team member should turn 90° to the right and keep looking straight ahead.
   (a) Are you able to see your team member?
   (b) Say 'Good morning' again.

(c) Did this seem louder, quieter or the same as before?

6. **Step 5:** Again, each team member should turn 90° to the right and keep looking straight ahead.
   (a) Are you able to now see your team member?

   (b) Say 'Good morning' again.
   (c) Did this seem louder, quieter or the same as before?

7. What can you say about sound waves versus light waves after this activity?

8.4 Activity-to-do: Doppler effect

Materials needed:

Tuning forks tied-up with a string, rubber mallet, stopwatch.

Background information:

Doppler effect, observed with all types of waves, is the increase (or decrease) in the perceived frequency of waves as the source and observer move towards (or away from) each other. When a sound-producing object moves towards the observer, the waves emitted by a source traveling towards an observer get compressed and the frequency of the sound waves increases, leading to a perception of a higher pitch.

Please note that Doppler effect does not happen because of an actual change in the frequency of the source. The source emits waves at the same frequency (f_{noise}) all the time, however, the observer perceives a different frequency (f_{L}) because of the relative motion between the source and the observer. Doppler effect observed for a sound source and listener is shown in figure 8.8, and the various situations are discussed below. In all cases discussed below, v_{noise} is the speed of the source, v_{L} is the speed of the listener, f_{noise} is the frequency at which the source emits, f_{L} is the perceived frequency of the listener and v_{sound} is the speed of sound in air.

1. **Case 1: Source moves towards a stationary listener**. If the source is moving towards a stationary listener, the listener perceives sound waves reaching him or her at a faster rate and $f_{\text{L}} > f_{\text{noise}}$ and perceived frequency is,

$$f_{\text{L}} = f_{\text{noise}}\left(\frac{v_{\text{sound}}}{v_{\text{sound}} - v_{\text{noise}}}\right) \tag{8.2}$$

2. **Case 2: Source moves away from a stationary listener**. If the source is moving away from a stationary listener, the listener perceives sound waves reaching them at a lower frequency ($f_{\text{L}} < f_{\text{noise}}$) and perceived frequency is,

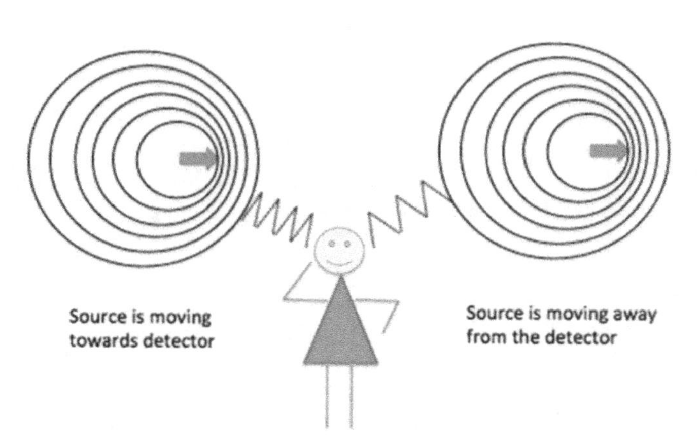

Figure 8.8. Doppler effect.

$$f_L = f_{\text{noise}}\left(\frac{v_{\text{sound}}}{v_{\text{sound}} + v_{\text{noise}}}\right) \tag{8.3}$$

3. **Case 3: Listener moves towards a stationary source**. When the listener moves towards a stationary source they '*meet*' the waves and the perceived frequency will be given as,

$$f_L = f_{\text{noise}}\left(\frac{v_{\text{sound}} + v_{\text{listener}}}{v_{\text{sound}}}\right) \tag{8.4}$$

4. **Case 4: Listener moves away from a stationary source**. If the listener is moving away from the stationary source, they will count fewer waves as they '*leave*' the source and the perceived frequency will be given as,

$$f_L = f_{\text{noise}}\left(\frac{v_{\text{sound}} - v_{\text{listener}}}{v_{\text{sound}}}\right) \tag{8.5}$$

5. **Case 5: Both source and listener move simultaneously** When both source and listener move simultaneously, we have to combine equations (8.2), (8.3), (8.4), (8.5) together to determine the perceived sound frequency as,
   (a) When source and listener move towards each other:

$$f_L = f_{\text{noise}}\left(\frac{v_{\text{sound}} + v_{\text{listener}}}{v_{\text{sound}} - v_{\text{noise}}}\right) \tag{8.6}$$

   (b) When source and listener move away from each other:

$$f_L = f_{\text{noise}}\left(\frac{v_{\text{sound}} - v_{\text{listener}}}{v_{\text{sound}} + v_{\text{noise}}}\right) \tag{8.7}$$

Example 8.2. While you are stopped at a red light, an ambulance producing a 3000 Hz tone moves toward you with a velocity of 50 m s^{-1}, what is the sound frequency you hear? What is the shift in frequency? Use the speed of sound at room temperature (= 344 m s^{-1}).

Solution:

Example 8.3. Suppose you are going down the road at 25 m s^{-1} toward a police cruiser producing a 2000 Hz tone and simultaneously the police cruiser is moving toward you with a velocity of 30 m s^{-1}, what is the perceived frequency as the police cruiser passes you?

Solution:

Experiment-to-do:

1. **Step 1: Swing the tuning fork**
   (a) Attach the tuning fork with a string.
   (b) Pull the sting taut and use a tape to mark at least three different lengths.
   (c) Note the three lengths here:
      1. $r_1 =$ ____________
      2. $r_2 =$ ____________
      3. $r_3 =$ ____________
   (d) Pick one mark of the tape and practice swinging the assembly at a steady constant rate.
   (e) Swing the tuning fork at that constant rate at each length (r) and record time taken for ten complete cycles in table 8.2.

Table 8.2. Average time for 10 swings of the tuning fork.

Length of string	Frequency of tuning fork (f_0)	No. of swings	Total measured time: Time 1 (t_1)	Time 2 (t_2)	Time 3 (t_3)	Average time for 10 swings (t_{av})
1. $r_1 =$		10				
2. $r_2 =$		10				
3. $r_3 =$		10				

2. **Step 2: Calculate the speed of source** (v_{noise})
   (a) The formula for speed is:

$$\text{Speed } (v_{noise}) = \frac{\text{Distance}}{\text{time}} \tag{8.8}$$

   (b) Find the distance (d) for 10 revolutions $=10 \times (2\pi r)$, use the value of r from Step 1 (c).
      1. $d_1 =$ ____________
      2. $d_2 =$ ____________
      3. $d_3 =$ ____________
   (c) Calculate the speed of source (v_{noise}) (which equals the speed of swings), use equation (8.8) and the values of distances (from Step 2 (b)) and the average times from table 8.2, $v_{noise} =$ ....

3. **Step 3: Doppler effect calculations**
   (a) Mathematically, the Doppler shift may be described by the following equations:
      1. Source moving towards you: observed frequency (f'):

$$f' = f_{\text{noise}}\left(\frac{v_{\text{sound}}}{v_{\text{sound}} - v_{\text{noise}}}\right)$$

      2. Source moving away from you: observed frequency (f''):

$$f'' = f_{\text{noise}}\left(\frac{v_{\text{sound}}}{v_{\text{sound}} + v_{\text{noise}}}\right)$$

      where, f is source frequency in Hertz (Hz), f' is the perceived frequency (Hz), v_{noise} is the speed of source (from step 2) and v_{sound} is the speed of sound (= 344 m s^{-1}).
   (b) Calculate the expected perceived frequency of your tuning fork by using the equations above and the average speed from your earlier measurements. The calculated perceived frequency:
      i. When the tuning fork moves towards you, (f')?

      ii. When the tuning fork moves away from you, (f'')?

   (c) The human ear can hear sounds with frequencies in the range 20 Hz to 20 000 Hz. However, it is most sensitive to frequencies between 500 and 4000 Hz and can distinguish between sounds that are a few Hertz different. Should a human with normal hearing be able to detect the frequency shift you calculated?

Part IV

Sound reception

IOP Publishing

The Physics of Sound and Music, Volume 2
A complete course text (Lab manual)
Samya Bano Zain

Chapter 9

Sound power and sound intensity

9.1 Terms to know about sound intensity and relevant concepts

1. Pressure, work, energy, power
2. Stress and strain
3. Sound and the ear
4. Amplitude and intensity of sound waves
5. Decibels
6. Sound intensity level (SIL)
7. Sound intensity (I) w.r.t. sound intensity at threshold of hearing (I_0)
8. Sound power
9. Sound pressure level (SPL)
10. Sound power level (SWL)
11. Sound power versus sound pressure
12. Loudness and loudness level.

doi:10.1088/978-0-7503-6350-1ch9

9.2 Pressure, force, energy

9.2.1 Activity-to-do: pressure

Materials needed:

Basketball; one 2-meter stick (or two 1-meter sticks); pressure gauge; ball air pump; clear tape; masking or painters tape; recording device (phone); thick (sharpie) marker.

Background information:

Pressure is defined as the amount of force acting upon an area.

$$\text{Pressure} = \frac{\text{Force}}{\text{Area}} = \frac{\text{N}}{\text{m}^2} = \text{pascal (Pa)}. \tag{9.1}$$

A large amount of pressure can be created by increasing the amount of force or by exerting the force over a smaller area (or by doing both). This means that the magnitude of the force isn't everything, sometimes you can have a larger impact by decreasing the surface of impact. This is why the tip of a nail is made pointy at one end to maximize the pressure exerted by the force applied by the hammer.

Example 9.1.

1. Suppose a block that weighs 100 N has a volume (2 m × 3 m × 0.5 m), what is the pressure exerted on the table if the block is set down on its side (2 m × 3 m)?

   Solution:

2. If the same block is now set down on its end which is 2 m × 0.5 m, what is the pressure?

   Solution:

We saw in chapter 4, activities 4.2 and 4.3. that a ball deforms more when dropped from 2 m than when it is dropped from 1 m. We know that a ball of mass (m) held at height (h) has potential energy ($=mgh$). When this ball is dropped it gains kinetic energy ($=1/2mv^2$). When this ball collides with the floor, the gained kinetic energy at the cost of initial potential energy, goes into deforming the ball from its original round shape. In the inelastic collision the kinetic energy of the colliding object is not conserved, which implies that there is a loss of energy in the collision, which is one of the reasons why the basketball never bounces back to the initial drop height.

When the pressure of the basketball is changed, for example by releasing the air, the time of contract between the ground and the ball is increased. This allows the basketball to compress more, which leads to a larger amount of kinetic energy of the basketball to transfer to the ground and hence decreases the bounce height.

Experiment-to-do:

In this experiment we test how the pressure of a basketball affects its bounce height when dropped.

1. Tape the 2 m stick to the wall with the masking tape.
2. Using the ball air pump, pump the basketball to full capacity. Measure the pressure with the pressure gauge.

   Initial pressure (max) = ___________ psi.

   Note: 1 psi (pound per square inch) is measured as one pound of force applied per one square inch. 1 psi is approximately equal to 6895 Pa.
3. Stand on a chair and drop the ball from 2 m.
4. Video the drop from initial drop height up to the second bounce of the ball. Replay the video in slow motion and record the max height after the first bounce in table 9.1.
5. Use the pressure gauge to release some of the air pressure from the ball. Record the new pressure reading in table 9.1. Step 2.
6. Keep releasing pressure and repeating the experiment until the ball no longer bounces.

Table 9.1. Height of bounce versus ball air pressure.

	Ball air pressure in		Initial drop height	Height of bounce			Average bounce height
	(psi)	(Pa)	(m)	Trial 1 (m)	Trial 2 (m)	Trial 3 (m)	(m)
1 =							
2 =							
3 =							
4 =							
5 =							

Analysis and discussion questions:

1. Use Excel to plot height of bounce (y-axis) versus ball pressure (x-axis). What does the graph look like? Is it a straight line?

2. Include a trend line on your Excel graph and display the equation on the chart. Write your equation here.

3. What is the value of the slope? What does the slope represent?

4. What could be your possible errors?

5. **Additional experimentation:** Rerun the experiment but this time change the surface on which the ball is bounced. Some surfaces to try could include, grass, concrete, clay, or artificial turf. What differences do you observe with each surface?

9.3 Stress and strain

9.3.1 Activity-to-do: strain

Materials needed:

Two different kinds of rubber bands (with different elasticities); scissors; clamp, various masses; meter stick.

Background information:

The effect of an applied force on an object is to either accelerate in the direction of the applied force or to change the shape of the object at the contact surface. This change can be permanent or temporary and is called '*deformation*'. When the change is temporary the object is called an '*elastic object*'. The fractional change in length is called the strain and is mathematically defined as,

$$\text{Strain} = \frac{\Delta L}{L} \tag{9.2}$$

where, L is the original length of the object under consideration and ΔL is the change in the length due to an external force. Note: strain is a ratio of lengths and hence it is a dimensionless quantity.

Experiment-to-do:

To test the effect of a tensile force applied to an object (figure 9.1).

1. Cut the rubber bands in half to have a long stretch.

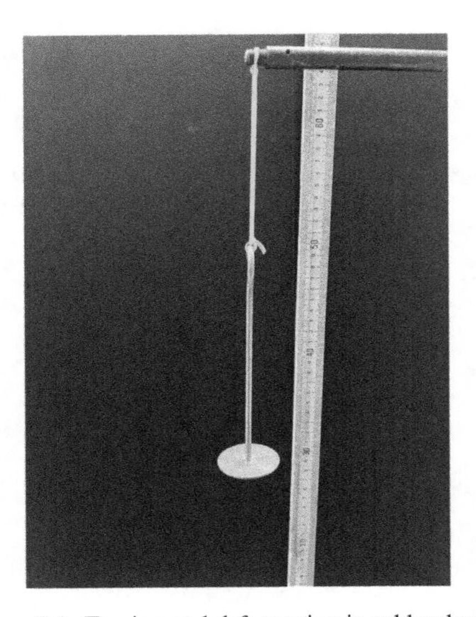

Figure 9.1. Tension and deformation in rubber bands.

2. Suspend one end of the rubber band from the clap.
3. Measure the length, this is the equilibrium position (no force added), call it initial length (l_0)
4. Attach the other end of the rubber band with a mass hanger. Usual mass hangers are 50 gms.
5. measure the new length (l_1).
6. Find change in length $\Delta l (= l_1 - l_0)$.
7. Record in table 9.2.
8. Calculate strain using equation (9.2)

Table 9.2. Stress and strain–1.

	Mass (kg)	Force (=mg)	l_0 (m)	l_1 (m)	Δl (m)	Strain
1						
2						
3						
4						
5						
6						

9.3.2 Activity-to-do: Hooke's law—stress and strain

Materials needed:

At least two different springs with different k values; stand; meter stick; various masses.

Background information:

Stress is the force between two objects that acts parallel to their interface. Stress is measured as force per unit area, mathematically given as:

$$\text{Stress} = \frac{\text{Force}}{\text{Area}} \tag{9.3}$$

The SI Units of stress are newton per meter squared (N m^{-2}) or pascal (Pa). When the stress on an object exceeds the *elastic limit*, the object will not return to its original length and the object will fracture. This is called the '*breaking point*'. The ultimate strength of a material is the maximum stress that it can withstand before breaking. The ratio of maximum load to the original cross-sectional area is called *tensile strength*.

Example 9.2. Anybody that does scuba diving knows that the pressure increases as they dive to greater depths. The increasing water pressure with depth limits how deep a submarine can go. Can you explain the sizes of the bubbles in figure 9.2?

Figure 9.2. Scuba diver.

Experiment-To-Do:

Investigate the relation between Hooke's law ($F = -k\Delta x$) and stress ($= \frac{F}{A}$).

1. Suspend one end of spring 1 from a clap.
2. Measure the length, this is the equilibrium position (no force added), call it initial length (l_0).
   Initial length (l_0) = ______________.
3. Suspend a mass hanger from the other end of the spring. Usual mass hangers are 50 gms.
4. Measure the new length (l_1).
5. Find change in length $\Delta l(=l_1 - l_0)$.
6. Record in table 9.3.

Table 9.3. Stress and strain–2.

	Spring 1			Spring 2		
	Mass (kg)	Force ($= mg$) (N)	Δl (m)	Mass (kg)	Force ($= mg$) (N)	Δl (m)
1.						
2.						
3.						
4.						
5.						
6.						

Analysis & discussion questions:

1. Use excel to plot force ($= mg$) versus displacement (Δl) for each spring. What does the graph look like?
2. Include a trend line on your Excel graph and display the equation on the chart. Write the equation here.
3. What is the value of the slope and what does it represent?
4. How does the behavior of the springs change as the value of spring constant (k) increases?

5. Relate Hooke's law to the stress–strain relationship. Start from Hooke's law:

$$F = -k(\Delta l)$$

Divide both sides by A:

$$\frac{F}{A} = -\frac{k}{A}\left(\Delta l\right)$$

Use the equation:

$$\text{Stress} = E\,\frac{(\Delta l)}{l_0}$$

9.3.3 Activity-to-do: stress–strain curve

Materials needed:

Two or three different kinds of materials with different elastic properties (rubber bands, thin strips of latex caulk; small-width electrical tape(<1 cm), fishing line (2 pound test), thin strips of plastic or garbage bags; boiled, dried and cooled noddles); safety glasses; stand; clamps; various masses; masking tape; meter stick; vernier calipers or micrometer (for measuring low elasticity elongations); some shock absorbing material (insulation materials work well).

Experiment-to-do:

Find the relationship between stress and strain for a given material.

1. Pick thin strips of 2–3 materials with varying elastic properties from the list given above, one very flexible, one medium and one completely non-flexible.
2. Measure the length of the strip.
3. Find the center and mark it with a marker.
4. From the center mark 3 cm above and below the center for high elasticity material for a total of 6 cm. For low elasticity materials, mark a distance of 5 cm above and below the middle for a total of 10 cm.
5. Measure the thickness and width of the material between the marks (use a vernier caliper or micrometer to measure the thickness for very thin materials)
6. Please put on your safety googles at this time and place cushioning to avoid damage to the table and/or floor.
7. Hang each material by clamps on one side and on the other side attach masses; via a mass hanger (50 gms) or masking tape (as required).
8. Now add more weights (in very small increments) very carefully and gently.
9. Measure the total distance between the marks and calculate the elongation (Δl) as each mass is added. Note: for low elasticity materials, there will be very little elongation.
10. Record the readings in table 9.4.

Table 9.4. Stress–strain curve, materials 1 and 2.

	Material 1					Material 2				
	l_0 =					l_0 =				
	A_0 =					A_0 =				
	Mass	Force	Stress	l_1 (m)	Strain	Mass	Force	Stress	l_1 (m)	Strain
1										
2										
3										
4										
5										
6										

11. Calculate the fractional change in length (strain = $\frac{\Delta l}{l}$) and the applied force (mg) that caused the strain.
12. Increase the applied force gradually, in steps, until the material breaks.
13. Note: if your material continues to stretch without additional weights being added, you may have to use a limit of time. Usually 30 s works well. After 30 s, one team member will need to support the masses and add more before again waiting for 30 s.
14. Continue adding weights until the material breaks (table 9.5).

Table 9.5. Stress–strain curve, materials 3 and 4.

	Material 3					Material 4				
	$l_0 =$					$l_0 =$				
	$A_0 =$					$A_0 =$				
	Mass	Force	Stress	l_1 (m)	Strain	Mass	Force	Stress	l_1 (m)	Strain
1										
2										
3										
4										
5										
6										

Analysis and discussion questions

1. Use Excel to plot a graph between the stress (magnitude of applied force per unit area) and the strain (elongation) for each material. Stress on y-axis and strain on x-axis.

2. Which sample had the greatest elongation? Which sample had the least elongation?

3. Calculate the elastic modulus (Young's modulus) for each material by finding the slope of the graphs before breaking point.

4. What is the relationship between stress and strain?

9.4 Sound pressure level, sound power level and sound intensity level

9.4.1 Sound pressure level, sound power level

1. **Sound pressure:** sound energy radiates outwards from the sound source as alternating compressions and rarefactions over an increasing area like an expanding balloon. These compressions and rarefactions are temporary changes in the atmospheric pressure that are pushed through the medium. The magnitude of change in the local atmospheric pressure caused by the vibration of the sound source is called *sound pressure.*

   The SI unit of sound pressure is pascal (Pa) and it is mathematically given by force provided by the sound source divided by the surface area over which the force acts,

$$\text{Sound pressure (p)} = \frac{\text{Force (N)}}{\text{Area (m}^2\text{)}} \tag{9.4}$$

2. **Sound power:** sound power is the acoustical energy emitted by a sound source per unit of time. It is an absolute value not affected by the environment.
3. **Threshold of hearing:** the faintest sound a typical human ear can detect has an intensity of 10^{-12} W m^{-2}. This intensity corresponds to a pressure wave that displaces the molecules in the compression region by 0.3 billionth of atmospheric pressure! This faintest sound that the human ear can detect is known as the '*threshold of hearing*'.
4. **Decibels:** the range of intensities that the human ear can detect is very large, hence physicists generally use a scale based on multiples of 10 to measure intensities called a '*logarithmic scale*'. The scale for measuring sound intensity is called the '**decibel scale**' (abbreviated dB), named after Alexander Graham Bell. Decibel is a dimensionless unit since it is a ratio of two quantities under consideration.
5. **Sound pressure level:** sound pressure level, abbreviated as SPL given as,

$$\text{Sound pressure level (dB SPL)} = 20 \log \frac{p}{p_0} \tag{9.5}$$

   The threshold pressure amplitude is $p_0 = 2 \times 10^{-5}$ Pa or 20 μPa. Since the atmospheric pressure $p_{atm} = (1.01 \times 10^5$ Pa), hence the ear is able to detect pressures,

$$\frac{p_0}{p_{atm}} = \frac{2 \times 10^{-5}}{1.01 \times 10^5} = 2 \times 10^{-10} \tag{9.6}$$

   This means detectable pressure is 2×10^{-10} times smaller than p_{atm}.
6. **Measuring sound pressure levels:** sound pressure levels are generally measured by sound level meters (SLMs). These SLMs are made of microphone, amplifier and a meter that gives a reading in decibels (dB). Sound instruments measure only sound pressure which varies depending on the surroundings. Ear sensitivity varies with frequency. A low frequency sound at a

certain power does not seem as loud as a higher frequency sound of the identical power. To account for this difference, a weighting scale has been developed. Sound power levels adjusted by this specific weighting scale are called A-weighted. Sound power levels in eight octave bands are calculated to a single A-weighted sound power number (LWA), as shown in figure 9.3.

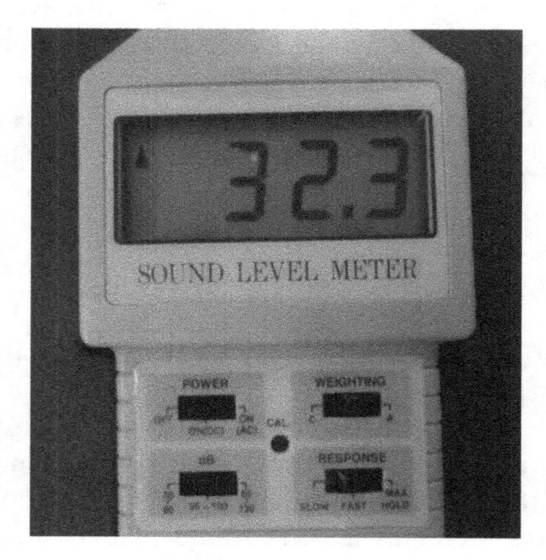

Figure 9.3. Sound level meter.

7. **Sound power level:** sound power level (SWL or L_w) is the acoustic energy emitted by a source which produces a sound pressure level at some distance. Sound power level (SWL) is a logarithmic measure of the power of a sound relative to a reference value.

$$\text{Sound power level (dB SWL)} = 10 \log \frac{W}{W_0} \tag{9.7}$$

where, W is the sound power measured in Watts and W_0 is the reference sound power, with commonly used value of 1×10^{-12} Watt.

SWL is very useful in quantifying the noise produced by a source. Hence it is used to predict the noise impact from a source without having to measure it. If you are asked to provide noise data for something, your acoustician will always appreciate data which is given as a sound power level (SWL).

Example 9.3. What sound pressure level corresponds to a sound pressure of 0.01 N m^{-2}?

Solution:

9.4.2 Intensity of sound waves

1. **Intensity of a wave:** the amount of energy that passes through a given area of any medium per unit of time is known as the '***intensity of a wave***'. The rate of energy transfer increases with increasing amplitude. For waves in three dimensions, the intensity is a measure of average power per unit area carried by the wave past a surface perpendicular to the wave's direction of propagation, mathematically expressed as,

$$\text{Intensity} = \frac{\text{Power}}{\text{Area}} \tag{9.8}$$

The SI units of the intensity of a sound wave are W m^{-2}. Intensity of a sound is an objective quantity and can be measured with instrumentation.

Example 9.4. A spherical sound source radiates at 25 W. What is the intensity of the sound wave 5 m from the source?

Solution:

2. **Intensity and distance from the wave source:** the intensity of sound waves decreases with increasing the distance from the source. s increases as shown in figure 9.4. For an isotropic source[1], the average power (energy per unit time) emitted is constant. Assuming that no energy is absorbed by the medium and there are no obstacles to reflect or absorb the sound, the intensity (I) at the distance r from an isotropic source is mathematically given as,

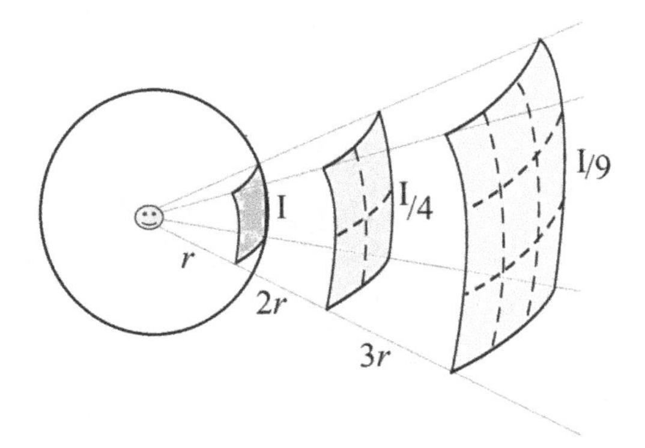

Figure 9.4. Expansion of 3D sound waves.

[1] An isotropic source is a point source that radiates uniformly in all directions. Hence, an isotropic sound source radiates sound uniformly in all directions.

$$\text{Intensity} = \frac{\text{Power}}{\text{Surface area}} = \frac{\text{P}}{4\pi \text{r}^2} = \frac{\text{Watt}}{\text{m}^2} \tag{9.9}$$

This is called the '**inverse square law**' and means that as the distance from the source is doubled, the intensity is quartered and energy passing through one square meter of surface on the sphere per second must drop off as $1/r^2$.

3. **Sound intensity level:** the sound intensity (measured in W m^{-2}) is a physical measure of sound wave amplitude, whereas sound intensity level (measured in dB) is its psychological measure. The equation that relates sound intensity to sound intensity level is:

$$\text{Sound intensity level, SIL (dB)} = 10 \log\left(\frac{I_2}{I_1}\right) \tag{9.10}$$

where SIL is the number of decibels and I_2 is the higher sound intensity being compared and I_1 is the lower sound intensity. For each power of ten increase or decrease in intensity of sound with respect to reference intensity, there is a (± 10) change in sound intensity level (in dB). Good reference numbers to remember are, 2 × Sound intensity = 3 dB and 10 × Sound intensity = 10 dB.

4. **Sounds just at the threshold of hearing.** Sounds just above the threshold of hearing, say at 1×10^{-12} W m^{-2},

$$\text{Sound intensity level (SIL)} = 10 \log \frac{\text{I}}{\text{I}_0} = 10 \log \frac{1 \times 10^{-12}}{1 \times 10^{-12}} = 10 \log(1) = 0 \text{ dB SIL}$$

Please note 0 dB does not mean that there is no sound, it only means that we just cannot hear it.

Example 9.5. Convert a measured sound intensity of 10^{-9} W m^{-2} to sound intensity level (dB SIL).

Solution:

5. **Sound intensity as a function of number of sources:** the total intensities of two sources is simply the sum of individual intensities of the sources. For example, if two violins are played equally loudly, the sound intensity at a certain point in space will be $2I$ as compared to intensity I for one violin. This means that there is a 3 dB increase in the SIL when we double the number of instruments. When we increase the number of instruments to ten, because of our logarithmic hearing, ten violins playing together will only sound twice as loud as one violin to us. For multiple sound sources the total SIL is,

$$\text{Total level} = \text{level of } a \text{ single source} + (10 \log(\text{number of sources})) \qquad (9.11)$$

Example 9.6. If one violin produces 70 dB SIL and we add a second violin also producing 70 dB SIL. What is the total sound intensity level?

Solution:

9.4.3 Activity-to-do: use the sound level meter to measure sounds

Materials needed:
SLM (physical or online); a 1 or 2 m stick.

Background information:
Use an SLM for measuring sound levels or you can use the online version if you do not have an SLM. You should familiarize yourself with the operation of the SLM by measuring the background sound level of the lab room. Turn on the meter by setting the range dial to 130 dB. Set the response of the meter to 'SLOW' and the weighting to 'C'. Measure the background sound level of the room by setting the weighting to C. Also, measure the background sound level using the A weighting of the meter.
Background sound level with C weighting = ____________.
Background sound level with A weighting = ____________.

Experiment-to-do:

1. Walk around campus and record various sounds, try to keep track of the distance you were from the noise maker.
2. Be back in 20 minutes!
3. Suggestions for things to record in table 9.6:
   - (a) Birds chirping.
   - (b) Running water.
   - (c) Ringing phone.
   - (d) Whistling.
   - (e) Closing a door, or a doorbell.
   - (f) Alarm clock.
   - (g) Regular speech, screaming, yelling, whispering.
   - (h) Musical instruments, piano, drum, bell, guitar etc.
   - (i) Air-conditioning or heating source in the room.

Table 9.6. Use the SLM to measure sounds.

	Distance away (m)	Thing recorded	Sound level (dB)
1.			
2.			
3.			
4.			
5.			
6.			
7.			
8.			
9.			
10.			

(j) Sing and speak the vowel sounds ah, ee and oo, slowly eliding between them.
(k) Sustain the consonant sounds ss and sh; f and v.
(l) Cars in the street. Please stay safe!
(m) Construction sites.
(n) Industrial activity.
(o) Basketball game.
(p) Music from clubs, concerts and/or festivals.

Discussion questions:

1. What were the loudest and the softest sound that you recorded?

2. What is the ratio of the loudest and the softest sound?

3. Should this ratio have units? If not, why not?

9.5 Loudness and loudness level

9.5.1 Activity-to-do: class activity: measure the sound level from various distances.

Materials needed:

SLM (physical or online); a constant amplitude noise maker (phone, stereo, drum, clapping etc); a 1 or 2 m stick.

Background information:

Intensity of a sound is an objective quantity that can be measured with sound meters. However, loudness is a psychological construct, which means loudness cannot be measured with a meter. However, we still use decibels (dB) as a unit for loudness.

1. **Loudness:** when you increase the volume of your stereo you are increasing the amplitude of the vibrations of molecules, which in turn the brain interprets as increasing the loudness of sound. Though not always true, in general we say that sounds with greater SIL or SPL will sound louder to us.

   Several odd anomalies exist in the mental representation of amplitude, such as:

   (a) Loudness is not addictive the way amplitudes are, rather loudness is logarithmic.
   (b) Loudness varies with the frequency of the incoming sound wave.
   (c) Loudness also varies with sound quality and ear sensitivity.
   (d) A human ear tends to amplify the sounds with frequencies in the range from 1000 Hz to 5000 Hz. This means that sounds with these frequencies will 'seem' louder to us. Hence, two sounds with the same intensity but different frequencies can be perceived to have different loudness.
2. **Loudness as a function of distance:** loudness is a subjective measure, and it falls off as the distance from the source increases. The perception of loudness also depends on the sound intensity, the frequency and duration of sound.

Experiment-to-do:

We will use an SLM for measuring sound levels or you can use the online version if you do not have an SLM.

1. Measure the sound level of a person talking from various distances.
2. Use the A-weighting and set the range dial to 130 dB for the sound meter and use FAST response to get peak sound levels to record data. This does not have to be quantitatively accurate.
3. Place the SLM at the 0 cm mark on the meter stick.
4. Start as close to the 0 cm mark and make the noise (using a noise maker or by clapping).
5. Read the SLM reading (dB) and record in table 9.7.
6. Make sure the sound is consistent throughout the experiment.
7. Move back 10 cm each time, make the same sound and record in the table.
8. Plot your data in Excel.

Table 9.7. Measure the sound level from various distances.

	Distance (m)	Sound level (dB)
1.		
2.		
3.		
4.		
5.		
6.		
7.		
8.		
9.		
10.		

Discussion questions:

1. What did you note from the excel data?

2. What role does distance play in the sound intensity? Explain your conclusions.

3. Based on the information found in the graph can you generalize your results in a mathematical formula?

IOP Publishing

The Physics of Sound and Music, Volume 2
A complete course text (Lab manual)
Samya Bano Zain

Chapter 10

The human factor

10.1 Terms to know about human hearing and sight and relevant concepts

1. Range of sight versus range of hearing
2. Speech production in humans
3. Signal processing
4. Bone conduction
5. Critical bands.

10.2 Hearing and sight

All electromagnetic radiation is called light, but we can only see a small portion, called the visible portion of light of this radiation. A typical human eye responds to wavelengths from about shortest of 390 (violet) to 750 nm (red) and in frequency band [400, 790] THz (THz = Tera Hertz = 10^{12} Hz). Human vision is remarkable, but the frequency range of vision does not even begin to compare to the frequency range of human hearing. Humans can hear frequencies as low as 20 Hz and as high as 20 000 Hz. Intensity ratio between sound that brings pain to our ears and the weakest sound is more than 10^{12} (1 000 000 000 000)!

Example 10.1. The frequency range that can normally be detected by the human ear is [20, 20 000] Hz. What is the corresponding range of wavelengths for sound in air at room temperature?

Solution:

doi:10.1088/978-0-7503-6350-1ch10

10.3 Left-brained versus right-brained

10.3.1 Activity-to-do: are you left-brained or right-brained?

Materials needed:

A small box with a few things inside (small pebbles or marbles); a ball (preferably a small one, like a tennis ball); empty tube (that of a kitchen towel will work fine); a note book paper with a small coin-sized hole cut in the middle; a sea shell.

Background information:

Sound signals received by the right ear nearly always get transmitted to the left side of the cerebral cortex and vice versa, signals falling on our left ear nearly always get transmitted to the right cerebral cortex. However, both sides of the cortex are connected and the information is processed throughout other parts of the brain. Scientists studying the brain over the years have evidence that suggests that for 97% of the human population, the left side is the dominant side and it is used for speech processing and is more analytical. The right side of the brain is the dormant/recessive (minor) side and it is specialized for nearly all non-speech related processing, for example, music, which requires holistic or synthetic processing. However, recognition of melodies requires both sides of the brain. It is experimentally found that melody recognition of non-musically inclined people is better if sound is heard in the left ear, whereas for the musically inclined melody recognition is better in their right ear. This suggests that musicians learn to process melodies in the dominant, analytical side of the brain. Sometimes trauma patients are able to sing songs learned before the onset of trauma, even if they are not able to speak the same words.

Experiment-to-do:

Working in teams of two to three, test yourself to see if you are left-dominant or right-dominant.

1. Work in teams of three students.
2. One team member will be the performer, one will be the questioner/enquirer and one will be the recorder. The questioner asks the questions, the performer carries out the tasks while the recorder should record the performer's responses in the table below.
3. **Hand tests:**
   (a) Write your name on a piece of paper.
   (b) Pick up the ball from the table.
   (c) Throw the ball to your partner.
4. **Foot tests:**
   (a) Drop the ball on the ground and gently kick it to your partner.
   (b) Take a couple of steps forward and jump off **one** leg.
   (c) Go to a stairway. Stand with both feet flat on the ground. Go up one step.
   (d) Come back to the landing, this time go one step down.
5. **Ear tests:**
   (a) Pick the sea shell and hold it up to your ear. Recorder should record which ear the performer held the sea shell up to.

   (b) Try to identify what is inside a box by putting an ear to the box. Recorder should record which ear the performer put the box up to.
   (c) Listen to a sound though a door. Recorder should record which ear the performer put to the door.
6. **Eye tests:**
   (a) Wink with one eye. Recorder should record which eye the performer winked with, left or right.
   (b) Look thorough the empty tube. Which eye did you look thorough?
   (c) Pick up the paper with the hole. Look at the clock on the wall. Move the paper closer and closer to your face while focussing on the clock. Which eye did you end at, left or right?

	Partner 1	Partner 2	Partner 3
Hand tests:			
Name on paper	Left/Right	Left/Right	Left/Right
Ball pick	Left/Right	Left/Right	Left/Right
Ball throw	Left/Right	Left/Right	Left/Right
Foot tests:			
Ball kick	Left/Right	Left/Right	Left/Right
Jump test	Left/Right	Left/Right	Left/Right
Step up	Left/Right	Left/Right	Left/Right
Step down	Left/Right	Left/Right	Left/Right
Ear tests:			
Sea shell	Left/Right	Left/Right	Left/Right
Box test	Left/Right	Left/Right	Left/Right
Listening through door	Left/Right	Left/Right	Left/Right
Eye tests:			
Wink test	Left/Right	Left/Right	Left/Right
Empty tube test	Left/Right	Left/Right	Left/Right
Paper test	Left/Right	Left/Right	Left/Right

Discussion questions:

1. So what side do you naturally favor?

2. Are you left-handed or right-handed?

3. Are you left-footed or right-footed?

4. Is your right eye dominant or is it your left?

5. What do you conclude, are you left- or right-dominant?

6. Are you left- or right-brained? Or do you use both sides of your brain?

10.3.2 Activity-to-do: class activity: calculate the percentage of the class who were left- or right- dominant

Compile the information from the entire class into one Excel sheet.

1. Calculate the percentage of class who were left- or right-dominant, by adding up the total number of students who mostly used their right hand instead of their left hand, and dividing this number by the total number tested.
2. You may want to keep the names of the students anonymous at this point.
3. Put all the voluntary information into the following table.

Domanant	Student 1 (Example)	Student 2	Student 3	···	% right	% left
Hand:	Left			···		
Foot:	Right			···		
Ear:	Left	···				
Eye:	Left			···		

4. Are more of your volunteers right-handed or left-handed?

5. What side is the most common and has the highest percentage in your subjects?

6. Calculate the percentage of people that are both same ear and eye dominant?

7. Calculate a percentage of people that are right-handed that are also right-footed dominant?

8. Are you more likely to be left-handed if one of your parents is left-handed?

9. What are some of the possible disadvantages for left-handed people (tools, writing materials etc)?

10. Do left-handed people have an advantage in sports?

11. Can you see a correlation?

10.3.3 Activity-to-do: bone conduction

Materials needed:
You; some form of audio recorder (computer; cellphone, etc).

Background information:
During speaking and singing, two different methods of hearing are utilized to transmit sound to the brain.

1. **Air conduction**: air conduction is the process by which an acoustic signal travels through the structures of the outer ear and middle ear and arrives at the cochlea.
2. **Bone conduction**: bone conduction is the process by which acoustical signals vibrate the bones of the skull to simulate the cochlea. The primary component of bone conduction is the inner ear.

   Bone conduction plays a very important role in the reception of high-intensity sounds and is the process by which acoustical signals vibrate the bones of the skull which simulates the cochlea. It plays a very important role in the reception of high-intensity sounds and the localization of sound. Underwater hearing is done entirely by bone conduction and is the reason why a person's voice sounds different to them when it is recorded and played back. Bone conduction tends to amplify the lower frequencies so most people perceive their own voice as being a lower pitch than others hear it. The sounds of humming or clicking one's teeth are heard almost entirely by bone conduction.

Experiment-to-do:

1. **Test 1:** First say this following common tongue twister out loud, then record yourself saying it and then listen back to the recording.

   Peter Piper picked a peck of pickled peppers;

   A peck of pickled peppers Peter Piper picked;

   If Peter Piper picked a peck of pickled peppers, where's the peck of pickled peppers Peter Piper picked.

   (a) Do you sound different on the recording?

   (b) Is the recording higher or lower pitched than what you expected yourself to sound like?

2. **Test 2: Humming test**
   (a) Hum and keep track of how loud it seems to you.
   (b) Next, close your ears with your fingers, thus interfering with air path, and hum with the same loudness as before. Does it sound louder or softer than before?

3. **Test 3: Clicking teeth**
   (a) Click your teeth and keep track of how loud it seems to you.
   (b) Next, close your ears with your fingers, thus interfering with air path, and click your teeth again. Does it sound louder or softer than before?

10.4 Activity-to-do: critical bands

Materials needed:

A piece of cloth or a scarf (elastic material will work better but is not necessary); a tennis ball; a ping-pong ball; a heavier ball (golf ball etc); a few (6–7) volunteers.

Background information:

The term critical band was first used by Harvey Fletcher in the 1940s and was later developed by Georg von Bekesy in the 1960s. Because of the basic natural structure of the Basilar membrane, different incoming frequencies resonate at different points along it. However, the Basilar membrane is a continuous element, it is generally incapable of resolving and analyzing inputs whose frequency difference is smaller than a certain limit. This limit is referred to as the '*critical band*'. Experiments carried out over the years have concluded that the average length of the critical band is ≈ 1 mm. Critical bands may be defined in two different ways:

1. **Definition 1: In terms of incoming sound:** when two pure tones are so close in frequency that there is considerable overlap in their amplitude envelopes on the Basilar membrane, they are said to lie within the same critical band.
2. **Definition 2: In terms of the structure of the ear:** critical band refers to the specific area on the Basilar membrane that goes into vibration in resonance with an incoming simple tone. Its length is determined by the elastic properties of the Basilar membrane (≈1.2 mm). There are about 24 critical bands along the Basilar membrane. Each critical band is about 1.3 mm long and is composed of about 1300 neurons.

Experiment-to-do:

1. Have the volunteers stand in a straight line.
2. Place the long cloth over the outstretched arms of the volunteers, leaving a bit of lag in between the arms of the volunteers.
3. Pick one side to be low-frequency side and the other end be the high-frequency end.
4. Divide each interval made by the arms to some frequency range. Make sure that volunteers are told the frequency ranges between their arms and the frequency range to the left and right side of their arms Example, [0, 50] Hz, [50, 100] Hz, [100, 150]....
5. The volunteers must remember the frequency ranges; to the right, middle and to the left of their arms
6. Have the volunteers close their eyes.
7. Put the tennis ball, ping-pong ball and the heavier ball at a few different points, at the middle and in between the volunteer's arms.
8. The volunteers will speak out loud the frequency ranges they think the balls are landing in.
9. Recorders record the volunteers responses and also note when the ping-pong ball is placed but the volunteers do not speak out.
10. Once all the tests are performed, volunteers may open their eyes.
11. The recorder will report their observations.

12. How many of the heavier balls were correctly predicted by the volunteers?

13. How many of the tennis balls were correctly predicted by the volunteers?

14. How many of the ping-pong balls were correctly predicted by the volunteers?

15. What do you conclude from your observations?

16. If you think of the heavier balls as having higher amplitude and the ping-pong balls as having lower amplitudes, what can you conclude from the observations above?

17. Can you see the issue with the naturally connected length of the critical bands and the interpretation of incoming frequencies?

IOP Publishing

The Physics of Sound and Music, Volume 2
A complete course text (Lab manual)
Samya Bano Zain

Chapter 11

Psychoacoustics

11.1 Terms to know about human hearing and sight and relevant concepts

1. Binaural hearing
2. Spatial hearing or localization
3. Echolocation
4. Masking
5. Adaptation and fatigue
6. Selectivity.

doi:10.1088/978-0-7503-6350-1ch11

11.2 Binaural hearing

11.2.1 Activity-to-do: binaural hearing

Materials needed:

1 paper towel tube; 2 wooden rulers or wooden sticks.

Background information:

Binaural hearing is the ability of humans to integrate information the human brain receives from the two ears. Just as Nature has provided us with the use of two eyes so we can see in three dimensions, Nature has also provided us with two ears so we can add a dimension to our hearing ability. Hearing in humans has two main purposes: communication (sound recognition) and warning (sound localization).

Binaural hearing is sometimes also called 'spacial hearing'. Experimentation has found that sounds entering the left ear are processed in the right hemisphere of the brain, while sounds entering the right side are processed in the left hemisphere. Then, both hemispheres work together to process and interpret the sounds. In this way, spacial hearing allows humans to estimate the location of sounds and the relative number of objects in the surroundings, which is essential for sound localization as a warning mechanism. Binaural hearing additionally allows us to separate a single voice from background noises

Experiment-to-do:

1. Work in teams of three.
2. Stand with your back to your partner, while one more person serves as the recorder. Put the end of the paper towel tube up to your left ear. Close your eyes and keep them closed. No peaking!
3. Have your partner stand about 4 to 5 feet behind you.
4. Have your partner tap the rulers together on the right side.
5. Then have your partner tap the rulers together on the left side.
6. Now have your partner tap the ruler a few times without you knowing where she or he is going to tap it. For each tap, you have to say which side the sound is coming from. Your partner will tap the rulers in these three places in any order:
   - (a) on the right side
   - (b) on the left side
   - (c) directly behind your head.
7. For each tap, the recorder places a mark for your response versus where the tapping actually was in table 11.1.

Table 11.1. Activity-to-do: binaural hearing.

	Tapped on			Heard on			Response	
	Left	Behind	Right	Left	Behind	Right	Correct	Not correct
1.								
2.								
3.								
4.								
5.								
6.								

8. Repeat this with the roles reversed and the recorder.
9. Compare your results, who got most correct answers?

10. Who had most trouble telling where the sounds coming from?

11. Discuss with your team members and explain your results.

11.2.2 Activity-to-do: spatial hearing or localization

Materials needed:

Ticking watch or metronome, you can use an online metronome (https://www.metronomeonline.com/).

Background information:

When multiple sounds reach our ears, they either interfere constructively or they interfere destructively with each other. For two individual sounds with amplitudes (A) that reach our ear the resultant amplitude reaching our ear can be anywhere in the range $[0, 2A]$. The apparent loudness of a sound to a receiver depends on the sound source and the distance between the receiver and the source. The loudness of a sound also depends on other factors, such as the direction of the wind, and obstacles or objects present between the source and the receiver.

Experiment-to-do:

1. Work in teams of three and take turns.
2. Seat team member 1 and blindfold them (or have them close their eyes).
3. Take a ticking watch or metronome at zero volume level and stand 3–4 feet in front of them.
4. Slowly increase the volume of the ticking watch or metronome and ask them to tell you when they hear 'ticking'. Record this volume level.
    1. Distance = __________
    2. Volume level = __________
5. Maintain this volume level for the rest of the experimentation for subject 1.
6. **Step 1:** Next, have subject 1 cover their right ear. Have team member 2 approach subject 1 from one side and record the distance when they say 'I can hear the ticking from the … right/left/front/back etc'.
7. Approach the blindfolded member from several different angles and record both the angles and the distance when they hear the ticking in figure 11.1.
8. **Step 2:** Repeat the same experiment with team member 1's left ear closed. Keep track of the angles and the distances in figure 11.2.
9. **Step 3:** Then repeat the same experiment with both of team member 1's ears opened. Keep track of the angles and the distances in figure 11.3.

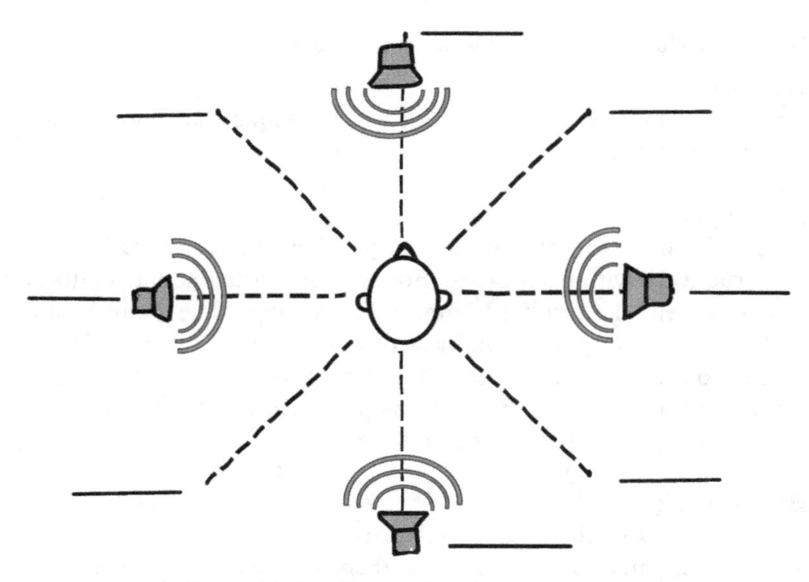

Figure 11.1. Localization. Step 1: right ear closed.

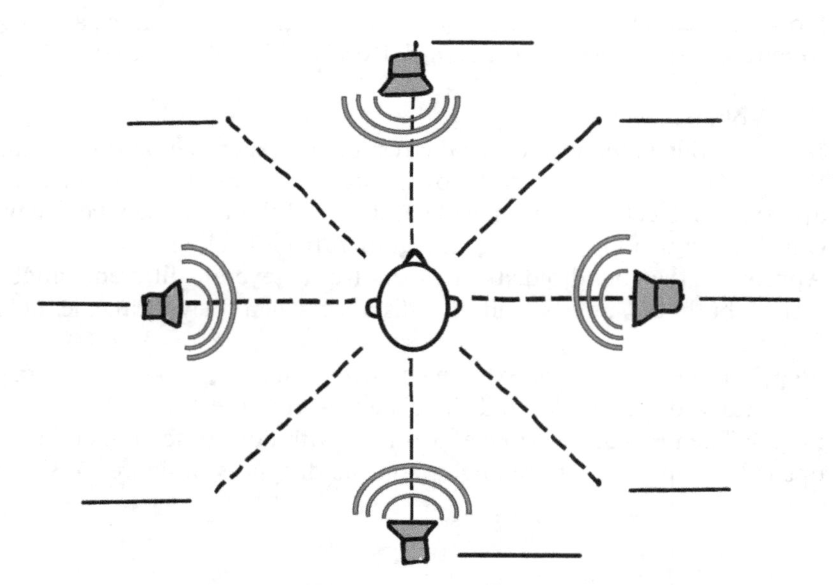

Figure 11.2. Localization. Step 2: left ear closed.

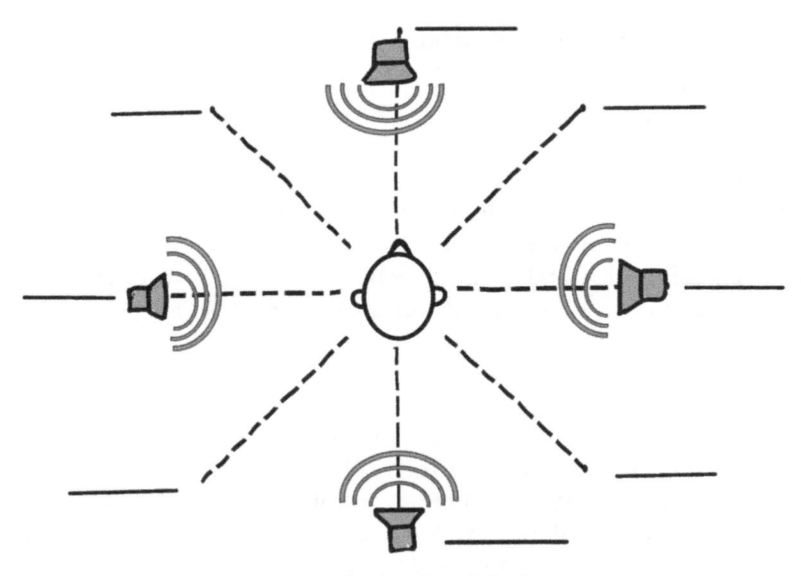

Figure 11.3. Localization. Step 3: both ears open.

Discussion questions:

1. Were your distances for both ears the same?

2. Were you better able to hear with one ear closed or both ears open? Explain.

3. What were your best and worst angles for detection?

11.3 Activity-to-do: echolocation

Materials needed:
2 rulers or wooden sticks.

Background information:
Echolocation is the ability developed by certain animals by which they determine the size and distances of objects in their surroundings by measuring the time delay between the emission of sound and any echoes that return from the environment. Bats emit ultrasound for echolocation and use echolocation to 'see' and hunt, most often in total darkness.

Experiment-to-do:
To 'see' without using your eyes—echolocation.

1. Work in teams of three: performer, observer and recorder.
2. One team member (you = observer) hold on to a chair and go around a couple of times and then sit on the chair. *Please keep your eyes closed throughout this activity and all three tasks.*
3. **Task 1:** team member 2 (performer), stands 7–10 feet behind the chair and claps the wooden sticks, while moving 5–10 regular steps either to the left or right.
4. You must tell if you hear the performer moving to the left or right.
5. Recorder should record here if the guesses match the observer guesses match the performer's actions.

6. Recorder, did the observer correctly describe the movements? If not, why not?

7. **Task 2:** next starting from the middle again (5–10 feet behind the chair); move randomly a few of steps to the left and right, clapping the sticks once per step.

8. Can you (observer) tell which direction the performer is moving?

9. Recorder should record the movements and the observations here.

10. Recorder, did the observer correctly describe the movements? If not, why not?

11. **Task 3:** next starting from the middle again (5–10 feet behind the chair); move in a circle (clock-wise or counter clock-wise) until you arrive at the front of the chair, clapping the sticks once per step.
12. Recorder should record the movements and the observations here.

13. Recorder, did the observer correctly describe the movements? If not, why not?

14. Once all tests are performed, the observer may open their eyes.
15. Discuss the results of the tasks. What do you observe? How well did you 'see' without using your eyes?

16. Each student in a team switch places and repeats the procedures.

11.4 Activity-to-do: masking

Materials needed:
Indoor and access to outdoors.

Background information:
The phenomenon by which one sound is obscured by the presence of another sound is called '*masking*'. Masking plays a very important role in the perception of combinations of tones. It allows humans to reject unwanted sounds, for example, the sounds of the heating, ventilation and air conditioning as a background sound in the classroom or theater. Usually the masking tone is fixed in frequency, whereas the masked tone has variable frequency.

Experiment-to-do:

1. **What is being masked?**
   (a) In the quiet classroom, close your eyes, concentrate on the sounds around you. Write down everything you can hear.

   (b) Next, go outside and find a nice place to sit. Close your eyes, concentrate on the sounds around you and write down everything you can hear.

IOP Publishing

The Physics of Sound and Music, Volume 2
A complete course text (Lab manual)
Samya Bano Zain

Chapter 12

Acoustics of rooms

12.1 Terms to know about room acoustics

1. Sound propagation; indoors versus outdoors
2. Direct sound and reflected sound
3. Precedence effect
4. Reverberant sound
5. Acoustics of rooms; ideal versus real
6. Absorption of sound; in air; with materials
7. Designing spaces.

12.2 Sound propagation

Sound waves travel away from a source. Sound is a pressure wave and as it travels its pressure decreases as a function of $1/r$, where r is the physical distance away from the source. This means that when the distance from the source doubles, the pressure 'p' is halved. In terms of sound pressure level (SPL), sound pressure level decreases by 6 dB each time distance is doubled. Attaining a good acoustical design does not happen by accident, but spaces must be designed very carefully. Special care must be taken to the final purpose and utilization of the space. Home listening room requirements are different from performance rooms and music studios.

1. **Indoors:** sound travels at 344 m s^{-1} at normal temperature (20 °C), so it will take between 0.02–0.2 s for the sound to reach the receiver directly from the source. This is called the '**direct sound**'. However, sound waves travel only a short distance indoors before they get reflected from the walls, ceilings and other objects in the room. Upon the first reflection the sound returns back to the receiver with a time delay and is called an **echo**. Not all echos present in the room are detectable, a distinct echo will only be detectable at a minimum

doi:10.1088/978-0-7503-6350-1ch12

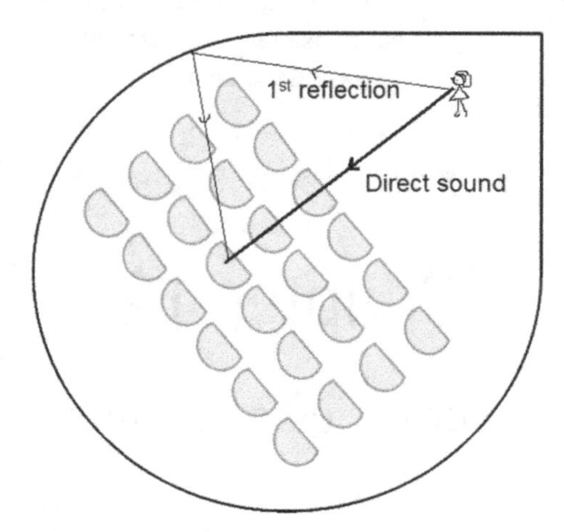

Figure 12.1. Sound reflections in a room.

50–100 ms after the previous sound. The first groups of reflections that reach the receiver are called '**early sound**' (figure 12.1).

2. **Precedence effect:** when the sound from a source reaches the listener directly, the level of it depends only on the physical distance between the source and the receiver. Humans can localize sound by determining the direction of the sound source. However, if sound reaches our ears from different directions our system gets confused, especially if the listener is far away from the source. Remarkably, our auditory processor continues to somehow figure out the direction of the source from the early sound reaching our ears. This remarkable ability of our auditory system is called '*precedence effect*'.
3. **Reverberant sound:** after the first reflections, the next reflections become smaller and closer over time and finally become merged together called '*reverberations*'. When the rate at which the source emits sound energy equals the rate at which sound energy is being absorbed by the environment, we call it reaching the '*reverberation level*'. Reverberant sound reinforces the direct sound and adds to the overall loudness of the sound. Reverberant sound is the characteristic of auditoriums that most people are familiar with. Our brains estimate the size of an enclosed space based on reverberation and echoes present in the signal that reaches us. Reverbs depend on frequency and will continue after the source stops emitting but decrease in amplitude until they get to inaudible amplitude and eventually stop.
4. **Reverberation time:** ee define '*reverberation time*' (RT) as the time required for the sound to '*fade away*' in an enclosed area after the source of the sound has stopped. The optimum reverberation time for any space depends on the performance requirements in the space. Spaces primarily used for speeches

need a shorter reverberation time so speech can be understood more clearly. However, if we make the reverberation time too short, we may negatively affect the loudness and tonal balance of sound. Reverberation effects are often used in music studios to add depth to musical sounds produced in the room.

5. **Reverberation time 60:** reverberation Time 60 (RT60) is used to when we want to make objective measurements of reverberation. RT60 is defined as the time it takes for the sound pressure level to reduce by 60 dB after the sound source stops. RT60 measures the time it takes for the loudest noise in a concert hall (100 dB) to fade to the background level (40 dB).
6. **Outdoor sound systems:** the considerations for an outdoor system are very different from indoor ones. Sound power is really important when we are talking about sound transmission in outdoor venues where we do not have reverberant sound field for reinforcement. For good sound transmission outdoors high efficiency, generally large strategically placed power amplifiers and arrays of loud speakers are needed.

12.3 Acoustics of rooms

1. **Acoustics of an ideal room:** an ideal room is defined as a space in which sound energy is uniformly distributed throughout. In a completely bare ideal room, all surfaces will absorb the same amount or fraction of energy and the reverberation time will be proportional to the ratio of volume to surface. If we have a hypothetical room with volume (V), having hard walls that do not absorb much sound and a window of area (A) through which sound may escape, then mathematically we can say,

$$\text{Reverberation time} = k\left(\frac{V}{A}\right) \tag{12.1}$$

where, k is a proportionality constant that depends on the dimensions of the room. When we use SI units for a room which absorbs all sound incident on it with volume V (m^3) and area A (m^2), k is calculated to be 0.161 s m^{-1}.
2. **Acoustics of a real room:** a real room has many surfaces that absorb sound and each has its own absorption rate. If we call the absorption coefficients for all the different surfaces a_1, a_2, ... having areas A_1, A_2, ... then the total absorption of the room is given by,

$$A_{\text{abs}} = a_1(A_1) + a_2(A_2) + \cdots. \tag{12.2}$$

The SI units for absorption (A_{abs}) are m^2, however, it is also sometimes expressed in **sabins** or **metric sabins**, named after Wallace Sabine, who did pioneering work in room acoustics. In general, one sabin is numerically equal to absorption of one square foot of an open window and one square sabin is numerically equal to absorption of one square meter of an open window.

3. **Absorption of sound in air:** air is an absorber of sound energy. Air absorption is negligible for frequencies less than 1000 Hz. However, if the room is very large, and high frequencies are present, air absorption cannot be neglected. Air absorption is also more pronounced in large auditorium settings. The absorption of air, in addition to other factors, depends on the temperature and relative humidity and for a large auditorium room it is given by,

$$\text{Reverberation time (RT)} = 0.161\left(\frac{V}{A + 4\,\text{mV}}\right) \tag{12.3}$$

where m is the air absorption constant or attenuation constant (SI unit = m^{-1}) that depends on the frequency of sound.

12.4 Designing spaces

When designing a space we have to take into consideration the primary use of the space. The requirements for various spaces are quite different from each other.

1. **A small room:** in a small room, the walls are close to each other and many reflections arrive from different sides within a few milliseconds of each other. In this case it is not important to differentiate between early sound and reverberant sound.
2. **Classrooms:** students must be able to clearly understand the instructor, who is usually talking at a sound level of about 46 dB for a student who is seated 30 feet away, if the background noise is at 36 dB.
3. **Home theaters:** one main consideration is the size and shape of the room. In general you want the room chosen to be a home theater to be not too large, in fact it should be an **acoustically small room**. Generally, a rectangular room with dimensions in the ratio (L: W: H:: 5: 3: 2) has been known to work the best. For example, a room with dimensions, $L = 15'$, $W = 9'$, $H = 6'$ would be acoustically good.
4. **Movie theaters:** the optimal place to experience the sound system in most theaters is to try to find a row two thirds the distance away from the front of the screen and a seat one or two seats from the exact center of the row.
5. **Concert halls:** among the world's greatest musical halls are those constructed in a rectangular shape, referred to as a '*shoebox*' design. Many modern concert halls have '*tuning features*', which include moveable reflecting panels and absorbing surfaces in the hall that may be changed, reordered, removed or added in order to customize the hall for different playing situations.
6. **Recording studios:** recording studios are very carefully designed, in order to keep the sound as close to the original as possible. This process involves very meticulous calculations of the correct physical dimensions of the rooms and proper soundproofing.

12.4.1 Activity-to-do: design a soundproof space

Materials needed:

A box (preferably a cardboard box); paper clips and painter's tape; sound making device (phone, stereo, etc); tone generator (manual or online) or a metronome (online metronome); sound-level meter; scissors; glue; miscellaneous materials (bubble wrap (cut into pieces); play-doh; parchment paper (cut into pieces); aluminum foil (cut into 10 cm × 15 cm pieces); sponges; various types of plates (Styrofoam, paper, plastic etc); felt (cut into 10 cm × 15 cm pieces); index cards; painter's tape; cotton balls; newspaper; computer paper; various types of straws (plastic, paper etc); craft sticks.

Experiment-to-do:

In this experiment we will see how to design the best soundproofing for 'our room'. Different material will be used to see if the decibel reading becomes lower (sound is absorbed) or if the decibel reading increases from the reference reading, that is, sound is not absorbed (or is reflected).

1. Place 'our room' (the box) on the table with the lid open, mark the outside of the box with masking/painter's tape.
2. Place the noise-making device (phone, stereo, etc) in the middle of the box.
3. Close the lid.
4. Place the sound-level meter outside the box at 20 cm away from the center. Mark this spot.
5. Turn on the meter by setting the range dial to 130 dB. Set the sound meter response to SLOW and the weighting to C. Measure the background sound level of the lab room. Also, measure the background sound level using the A-weighting of the meter. What is your initial decibel reading?
   Background sound level with C weighting = __________
   Background sound level with A-weighting = __________
6. It is extremely important that sound level measurements are made using the correct frequency weighting, hence leave your meter at the A-weighting, FAST response and set the range dial to the highest level (120 dB), for the rest of the lab.
7. First, test the box reading without any materials to get your initial reading. What is your initial decibel reading? Initial dB reading = __________
8. Choose three different materials that you hypothesize will be the best at soundproofing the room. Write the materials you chose here.
   Material 1 = __________
   Material 2 = __________
   Material 3 = __________
9. Choose three different materials that you hypothesize will be the least effective at soundproofing the room. Write the materials you chose here.
   Material 1 = __________
   Material 2 = __________
   Material 3 = __________

Table 12.1. Soundproofing data.

	Material used	Initial dB reading	Final dB reading	Change in dB
1				
2				
3				
4				
5				
6				

10. Make sure to use paper clips and/or tape to hold the materials around the box.
11. Without changing the placements for the box and the sound-level meter measure the dB reading again with the various materials placed one layer thick around the box. Once you have tested your materials, record in table 12.1.
12. Calculate the change in decibels by subtracting measured dB reading with respect to initial decibel reading.
13. Plot your data on a bar graph in Excel hearts, material on the x-axis and change in dB on the y-axis.
14. What do you observe in the graph?

Discussion questions:

1. Which material was the least effective at soundproofing the room?

2. Which material was the most effective at soundproofing the room?

3. Was your hypothesis valid?

4. Will you soundproof your entire house with the material you found best soundproofs your 'room'? If not, why not?

12.4.2 Activity-to-do: how much soundproofing material works best?

Materials needed:

Soundproofing material that worked best in activity 12.4.1.

Experiment-to-do:

In this experiment we will do a cost-to-benefit ratio analysis of the different materials that best soundproof 'our room' (the box) (figure 12.2).

1. Use the material that worked best in activity 12.4.1.
2. Hypothesize how many layers provide the best cost-to-benefit ratio.

3. Keeping the distance of the sound-level meter the same, change the number of layers of the material.
4. Find the initial dB reading for no layer situation, with just the box.
   initial dB reading = __________

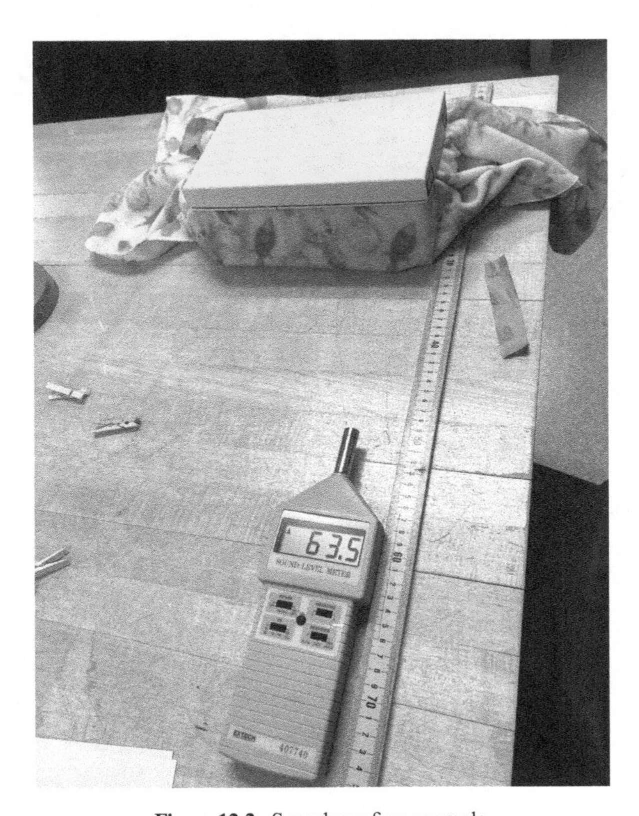

Figure 12.2. Soundproofing example.

Table 12.2. Soundproofing data.

Layers	Initial dB reading	Final dB reading	Change in dB	Total Cost	Cost per layer
0					
1					
2					
3					
4					
5					

5. Test with different layers of the material that worked best in activity 12.4.1. and record your data in table 12.2.
6. Change in decibels is calculated by subtracting measured dB minus the initial decibel reading.

Discussion questions:

1. What do you observe as you increase the number layers?

2. What is the best cost-to-benefit ratio?

3. Was your hypothesis valid?

Part V

Of sound and music

IOP Publishing

The Physics of Sound and Music, Volume 2
A complete course text (Lab manual)
Samya Bano Zain

Chapter 13

Musical tones, pitch, timbre and vibrato

13.1 Terms to know about musical tones and pitch

1. Musical tones and pitch
2. Vibrato
3. Amplitude modulation (AM)
4. Frequency modulation (FM)
5. Just notable difference (JND)
6. Timbre.

doi:10.1088/978-0-7503-6350-1ch13

13.2 Musical tones and pitch

A sound wave described by a single frequency is called a '*pure tone*'. Most sounds are not pure tones but a combination of many pure tones. Many sounds are a combination of frequencies that are harmonically related. One way to characterize such combinations is a property called '*pitch*'. Pitch like loudness is a subjective quality, it depends on how humans process sound. The American Standards Institute (1960) defines pitch as:

'*That attribute of the auditory sensory sensation in terms of which sounds may be ordered on a scale extending from low to high.*'

Pitch allows sounds to be ordered based on the frequency of the vibration of the sound-producing object, where '*high*' pitch means very rapid oscillations of the air molecules (high frequency) and '*low*' pitch means slower vibration of air molecules (low frequency). Because pitch is a subjective sensation, two people hearing the same sound may assign it a different position on a pitch scale. In fact, listeners may assign a different pitch to a sound depending upon whether they hear it from the left ear or the right ear. This is called '*binaural diplacusis*'. The basic unit for pitch in most musical scales is the '*octave*'. Notes judged an octave apart have frequencies nearly (but not always) in the ratio 2 : 1.

Some important terminologies that are important to know about pitch include:

1. **Definite pitch:** definite pitch is a sound or a musical note where it is possible or relatively easy to differentiate the pitch. Sounds with definite pitch have harmonic frequency spectra or close to harmonic spectra.
2. **Indefinite pitch:** Indefinite pitch is a sound or a musical note where it is impossible or extremely difficult to separate out a particular pitch from the multiple pitches produced. Sounds with indefinite pitch do not have harmonic spectra or have altered harmonic spectra a characteristic known as '*inharmonicity*'. Examples include percussion instruments that do not have a particular pitch.
3. **Concert pitch:** concert pitch is the pitch reference to which musical instruments are tuned for a performance. These days, A above middle C is usually set as the concert pitch with frequency = 440 Hz, and is often written as 'a4' or 'A = 440 Hz' or 'A440').
4. **Absolute pitch:** absolute pitch (AP), or perfect pitch, is the ability of certain humans to name or reproduce a tone without reference to an external standard.
5. **Relative pitch:** relative pitch is a skill, which can be learned through intense ear training and practicing. Many musicians have quite good relative pitch.
6. **Virtual pitch:** virtual pitch is the experimentally established phenomenon in which one's brain interprets tones in music that don't actually exist.

13.2.1 Activity-to-do: determine the relative or absolute pitch for the class members

Materials needed:

Piano; online pitch detector (physical or https://www.onlinemictest.com/tuners/pitch-detector/), class members.

Experiment-To-Do:

1. As a class, choose three notes. Write them down here.
   Note 1 = = __________
   Note 2 = = __________
   Note 3 = = __________
2. Play the three notes one at a time on a piano (example, the notes B, C, G).
3. Have the class practice singing each note.
4. Play the notes again and have the class members sing each note.
5. Ask the class members if they think they will be able to sing three notes on pitch after hearing them:

YES/NO

6. After the class has practiced three times, each member should sing each note.
7. Use the manual or online pitch detector to determine each class member's pitch.
8. Record in table 13.1.
   (a) If the class member is on pitch, flat, or sharp on each note. (If flat or sharp, write down by how much.)
   (b) Ask the class member if they thought that they had sung on pitch.
   (c) Record if there is a difference between the volunteers' opinion of note played and the actual note being played.

Table 13.1. The relative pitch for the class.

Note played	Volunteers' years of music lessons	Actual note played	According to pitch detector volunteer is			Volunteers' opinion of note played	Difference
			On pitch	Flat	Sharp	(Correct or not)	(Yes/No)
1.							
2.							
3.							

Discussion questions:

1. Was there a correlation between number of years of music lessons and singing accuracy?

2. What percentage of the class sang all three notes on pitch?

3. What percentage of the class knew whether they were on pitch or not?

4. Did most of the class sing more flat or more sharp or was it about even?

5. What could be the sources of errors in this study? Do you have enough volunteers? Are your results statistically significant?

13.2.2 Activity-to-do: pitch experiment—evaluate frequencies of tuning forks

Materials needed:

Oscilloscope; tuning fork; microphone; connector to connect microphone to oscilloscope.

Background information:

An oscilloscope is an electronic measuring instrument that creates a visible two-dimensional graph of one or more electrical potential differences. The horizontal axis of the display represents time and the vertical axis usually shows voltage. The display is caused by a 'spot' that periodically 'sweeps' the screen from left to right. The oscilloscope repeatedly draws a horizontal line called the trace across the middle of the screen from left to right. One of the controls, the timebase control, sets the speed at which the line is drawn, and is calibrated in seconds per division. The vertical control, sets the scale of the vertical deflection, and is calibrated in volts per division.

For a periodic input signal, nearly stable trace is obtained by setting the timebase to match the frequency of the input signal. For example, an input of 50 Hz sine wave has a period (1/50 =) 20 ms, so the timebase should be adjusted so that the time between successive horizontal sweeps is 20 ms. This mode is called continual sweep.

Experiment-to-do:

To experimentally find the frequencies of tuning forks.

1. Attach the microphone to the oscilloscope, as shown in figure 13.1 (left).
2. Set the oscilloscope as seen in figure 13.1 (right) and do the following:
   (a) Time: read off the oscilloscope dial (SEC/DIV); remember to convert from 'milli-s' to 's' as shown in figure 13.1 (right), in this case the time is 2 ms.
   (b) Read distance on the x-axis on the screen: read the wavelength in terms of number of boxes. 1 box = 1.0; each box has 5 divisions.
   (c) Then calculate period as = (time) × (distance on x-axis)
   (d) Evaluated frequency is then found by = 1/period.
   (e) **Solved example 2:** as shown in figure 13.2 (bottom).

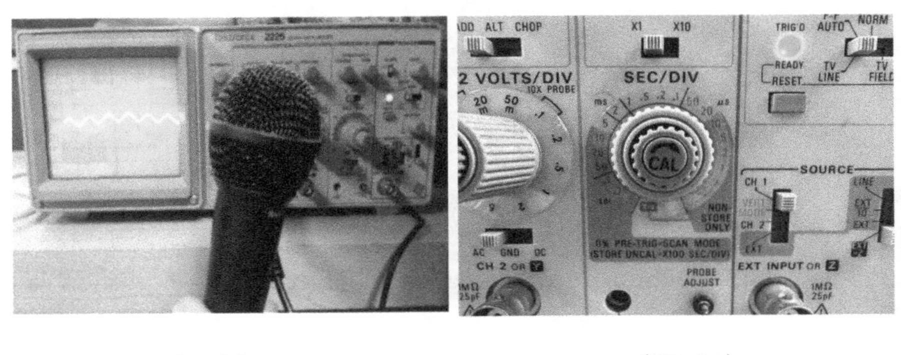

(Left) (Right)

Figure 13.1. (Left): oscilloscope setup. (Right): how to read the time on the oscilloscope for table 15.1, in this case the time is 2 ms.

Solved Example 1:

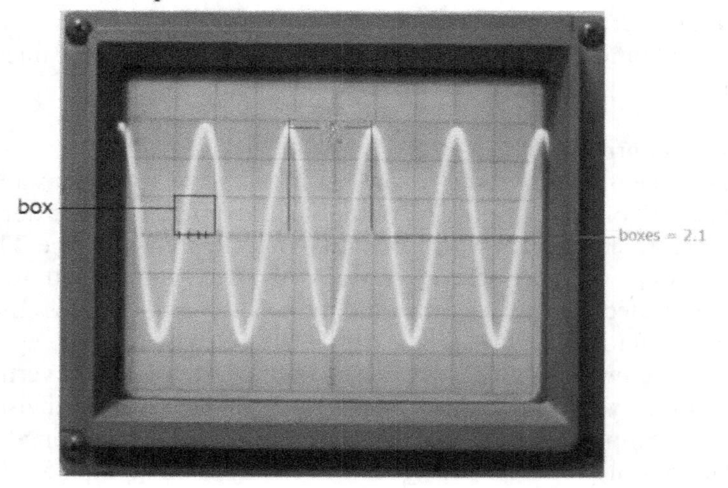

Solved Example 2:

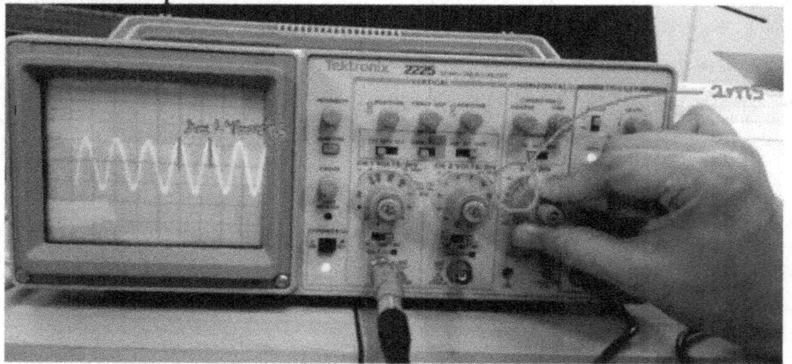

Figure 13.2. Solved examples with oscilloscope details.

Oscilloscope readings:

i. Time = 0.002 s.
ii. Wavelength (number of boxes on the x-axis) = 1.4.

Calculations:

i. Calculated period = 0.002 × 1.4 = 0.0028.
ii. Evaluated frequency of the tuning fork = 1/0.0028 = 357.14 Hz

3. Find the values for multiple tuning forks and record all values in table 13.2.

Table 13.2. Evaluating the frequencies of tuning forks.

	Oscilloscope readings						
	Time (ms)	Time (s)	Distance on x-axis	Calculated period (s)	Evaluated frequency (1/s = Hz)	Actual frequency (Hz)	% difference
Example 1	1	0.001	2.1	= 0.001 × 2.1 = 0.0021	476.2 Hz		
Example 2	2	0.002	1.4	= 0.002 × 1.4 = 0.0028	357.14 Hz		
1							
2							
3							
4							

Discussion questions:

1. What is the average % difference between evaluated and actual frequencies?

2. What could be your possible sources of errors?

13.3 Vibrato

The effective result of a sine wave of audible frequency (say 512 Hz) '*frequency modulated*' by another sine wave having frequency 1 Hz is the production of another sine wave, centered at the frequency of the original wave, where this wave produces a tone that varies slowly up and down in frequency, this effect is called '*vibrato*'. Vibrato adds a distinctive flavor to the tone of the melody. For example, the vibrato of a singer's voice adds significantly in distinguishing it from the sounds of other instruments. Vibrato is used to enhance musical performance.

Vibrato is also sometimes used interchangeably with the periodic changes in amplitudes. However, these periodic changes in amplitudes in a melody should be correctly referred to as *Tremolo*. The '*diaphragm vibrato*' of a flute player is close to pure tremolo.

1. **Amplitude modulation (AM):** amplitude modulated (AM) radio waves are used to carry commercial radio signals in the frequency range from [540–1600] kHz. A carrier wave, the basic frequency assigned to a particular radio station is varied or modulated in amplitude by an audio signal. The resulting wave has a constant frequency, but has varying amplitude that is a function of the information it carries, as seen in figure 13.3.
2. **Frequency modulation (FM):** frequency modulated (FM) radio waves are used for radio transmission in the frequency range of [88–108] MHz. In this case the a carrier wave is modulated in frequency by the audio signal, producing a wave of constant amplitude but varying frequency, as seen in figure 13.4.

Please note that AM without FM is possible, however, it is virtually impossible to have FM without having AM. This is because we cannot ever completely eliminate resonances in the instruments and the rooms the instruments are played in.

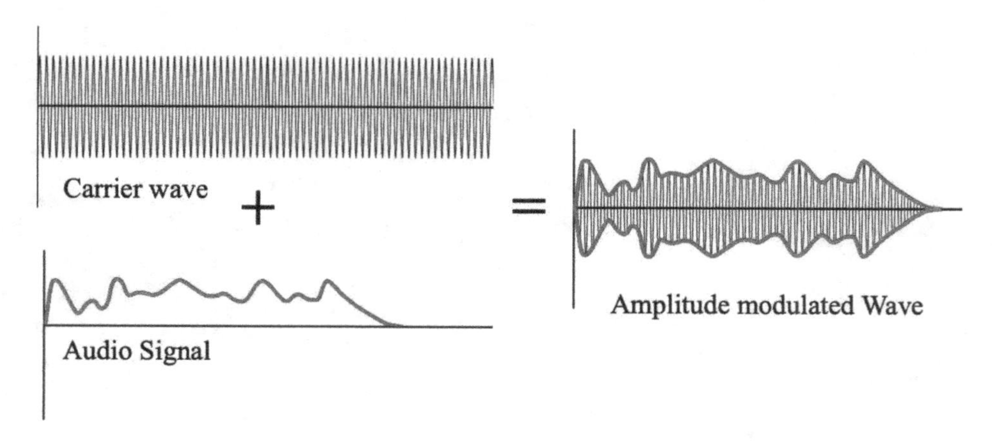

Figure 13.3. Amplitude modulation for AM radio. (a) A carrier wave at the station's basic frequency. (b) An audio signal at much lower audible frequencies. (c) The amplitude of the carrier is modulated by the audio signal without changing its basic frequency.

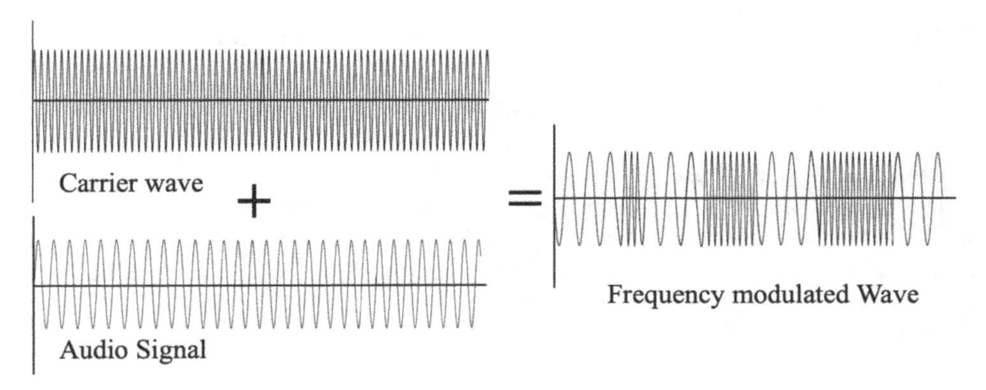

Figure 13.4. Frequency modulation for FM radio. (a) A carrier wave at the basic frequency of the station. (b) An audio signal at much lower audible frequencies. (c) The frequency of the carrier is modulated by the audio signal without changing its amplitude.

Example 13.1.

a. Does the frequency 1450 kHz correspond to AM or FM radio channel?
   Solution:

b. Does the frequency 100.1 MHz correspond to AM or FM radio channel?
   Solution:

c. Calculate the wavelengths of a 1450 kHz radio signal, a 100.1 MHz radio signal. [Hint: speed of light $c = 3 \times 10^8$ m s^{-1}]
   Solution:

d. Compare the wavelengths to the signals above, what does that tell you?
   Solution:

13.3.1 Activity-to-do: dependence of pitch on diameter and length

Materials needed:
Assorted test tubes, water, vernier calipers.

Background: how to use a vernier caliper
The vernier caliper, as shown in figure 13.5 (left), is a device used when extremely precise measurements are required. It is used in measuring length in order to gain an additional digit of accuracy compared to measurement done with the help of a simple ruler.

- Parts of the caliper:
  1. Outside jaws: used to measure external diameter or width of an object.
  2. Inside jaws: used to measure internal diameter of an object.
  3. Depth probe: used to measure depth of an object or length of a column.
  4. Main scale: gives measurements of up to one decimal place.
  5. Vernier scale: gives measurements of up to two decimal places.

Experiment-to-do:
To evaluate how the pitch depends on the diameter and the length of a material, in this case a test tube.

1. Measure the internal and external diameters and the length of the column for each test tube by using vernier calipers and include the values in table 13.3.
2. Gently blow across the test tubes one by one and note down the pitch (as, low, high, medium) in table 13.4. Hint: your lower lip and your chin must both touch the bottle.
3. Fill each test tube to about one-fourth and then one-half full of water. Gently blow across the test tubes one by one and note down the pitch (as, low, high,

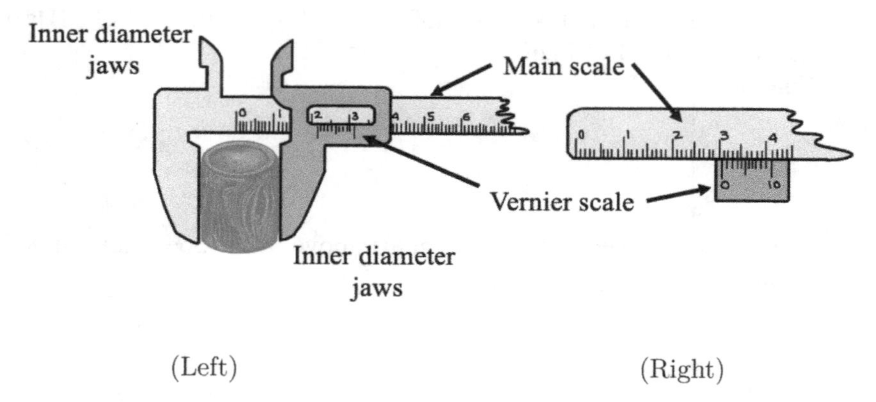

Figure 13.5. (Left) vernier caliper. (Right) vernier caliper scale reading = 3.03 cm. The 3.0 cm is read from the main scale, then vernier scale reading is read at the first point of closest alignment between the two scales. In this case, vernier scale reading is the third division as it aligns the best with the main scale. Hence the final reading of the vernier caliper is 3.03 cm.

Table 13.3. Dimensions of the test tubes.

	Outer diameter (cm)	Inner diameter (cm)	Length of the column (cm)
Test tube 1 =			
Test tube 2 =			
Test tube 3 =			
Test tube 4 =			
Test tube 5 =			

Table 13.4. Empty test tube measurements.

	low, medium, or high pitch
Test tube 1 =	
Test tube 2 =	
Test tube 3 =	
Test tube 4 =	
Test tube 5 =	

Table 13.5. Filled test tube measurements.

	Quarter-filled	Half-filled tube
Test tube 1 =		
Test tube 2 =		
Test tube 3 =		
Test tube 4 =		
Test tube 5 =		

medium) in table 13.5. Compare what happens to your perception of pitch from step 2.

Discussion Questions:

1. Which test tube had the highest note, or pitch? Which had the lowest pitch? How did the diameter and length of the column come into consideration?

2. See what happens when you try to measure the output by an oscilloscope?

3. What do you think would happen if you added more water to the test tube that is half full?

4. Test your prediction by adding more water. Was your prediction correct?

13.4 Activity-to-do: just notable difference (JND) for sound—frequency JND

Materials needed:

Tone generator with volume control; quiet room.

Background information:

Psychophysics is the study and measurement of subjective human experiences by examining how environmental stimuli affect humans. JND is a part of psychophysics and is the smallest 'frequency change' that humans can hear. Suppose a 440 Hz sine wave is '*frequency modulated*' by another sine wave. If we start both waves when there is no frequency difference between them and slowly increase the difference. At first we will not be able to tell the difference between the two sine waves, but if the frequency of the modulating wave is increased beyond a certain limit the pitch change will become noticeable. This particular frequency may vary from human to human and is called '*frequency just noticeable difference*' or '*frequency JND*'.

We can measure JND for nearly all frequencies in the audible range by varying the original frequency with respect to the modulating frequency. Experimentally, it has been found that frequency JND is between 0.5–0.6% over most of the audible range. However, this variation is larger for very high and very low frequencies. For most frequencies, the JND is much smaller than the critical band of the basilar membrane, which means that even though a large area of the basilar membrane vibrates in response to the frequency, humans can distinguish between changes in the frequency, generally 1/10th the width of the basilar membrane.

One way to experimentally determine JND is to compare a fixed reference sound with several test sounds. JND will be the smallest difference that is just audible between the test sound and the reference sounds. Multiple test subjects should also be used because human perception of sound varies with age. Using a relative scale, JND can be mathematically expressed as,

$$\mathrm{JND} = \left| \frac{\Delta f}{f_0} \right| \tag{13.1}$$

where, Δf is the absolute value of difference threshold and f_0 is the reference frequency. For two pure tones, JND is about 7% for low frequencies and about 15% for high frequencies. Most studies place it around 3% in the 100 Hz range, but only 0.5% in the 2000 Hz range. Musicians can tell the difference between two notes played one-half step apart and can perceive them to be two separate notes.

Example 13.2. Find the JND if the reference frequency is 3000 Hz, and the absolute change in signal frequency is 30 Hz.

Solution:

Experiment-to-do:

Calculate the JND in sound for the class.

1. Use the tone generator to generate a tone at 440 Hz.
2. Keep the volume at mid-range.
3. Select one team member and have them sit with their eyes closed.
4. Play the tone for 2 s at a particular frequency (say 440 or 512 Hz) and tell them to use this as a reference tone. You may not tell them the actual frequency of the tone. In this activity I will use 440 Hz as the reference frequency, but you may choose your frequency of choice.
5. Increase or decrease the frequency by 1 Hz and ask the team member if the second tone is higher or lower than the first tone.
6. Record answers in a table 13.6
7. Continue the same procedure, increasing or decreasing the frequency and always comparing them to the chosen reference frequency.
8. Note when the member can detect difference in frequency as compared to the reference frequency.
9. Make sure to test all members of the class and record the data.
10. **How to calculate JND:**
    (a) If the team member could not tell the difference between the reference and the played frequency ± 1 that member's frequency difference is '1' and their JND is 1/440 or 0.002.
    (b) If the team member cannot tell the difference between the reference and the played frequency ± 2 that member's frequency difference is '2' and their JND is 2/440 or 0.005, etc.
11. **Average JND for class:**
    Calculate the average JND for all class members and record them in table 13.7.
    JND Class average = __________

Table 13.6. JND—frequency.

	Frequency played		Subject answered Higher/lower	Correct/incorrect
1	435 Hz			
2	436 Hz			
3	437 Hz			
4	438 Hz			
5	439 Hz			
6	440 Hz	Reference	–	–
7	441 Hz			
8	442 Hz			
9	443 Hz			
10	444 Hz			
11	445 Hz			

Table 13.7. JND—frequency.

Person	Reference frequency (f_0)	Frequency difference Δf	JND $= \left\| \frac{\Delta f}{f_0} \right\|$	Class average = ________	Higher/lower than average
1					
2					
3					
4					
5					
6					
7					
8					
9					
10					
11					
12					
13					
14					

Discussion questions:

1. What do you find about the class average? Were you surprised?

2. What do you find about yourself with respect to the class average? Were you surprised?

IOP Publishing

The Physics of Sound and Music, Volume 2
A complete course text (Lab manual)
Samya Bano Zain

Chapter 14

A musician's graph paper and musical scales

14.1 Terms to know about musical graph paper

1. Logarithms
2. Musical staff
3. Musical scale
4. Tuning
5. Musical intervals
6. Important terms to know; intonation, tonic, consonance and dissonance, just intonation, temperament, false pattern recognition.

doi:10.1088/978-0-7503-6350-1ch14

Calculations with logarithms.

1. The logarithm to the base 10 of a number x is the power to which 10 must be raised in order to equal x.

   Example 1:
   $100 = 10^2$.
   Logarithm of 100 (to the base 10) is 2, or, $\log 100 = 2$.
   Example 2:
   $10\,000 = 10^4$.
   Logarithm of 10 000 (to the base 10) is 4, or, $\log 10\,000 = 4$.
2. Values of logarithms are given in table 14.1:
3. **Properties of logarithms.**
   Basic important logarithmic properties are given here.
   For given positive integers A, B and n,

   - **Product property:** $\log(AB) = \log A + \log B$
     Example:
     $\log(15) = \log(5 \times 3) = \log 5 + \log 3 = 0.699 + 0.477 = 1.176$.
   - **Quotient property:** $\log(\frac{A}{B}) = \log A - \log B$
     Example: $\log(\frac{5}{3}) = \log 5 - \log 3 = 0.699 - 0.477 = 0.222$.
   - **Power rule:** $\log A^n = n \log A$
     Example: $\log 10^3 = 3 \log 10 = 30$.

Table 14.1. Values of logarithms.

x	$\log(x)$
1	0
2	0.301
3	0.477
4	0.602
5	0.699
6	0.778
7	0.845
8	0.903
9	0.954
10	1.000

Example 14.1. Using the properties of logarithms and the values given in table, find:

1. $\log(400) =$

2. $\log(2.5) =$

3. $\log(125) =$

4. $\log(2.16) =$

Example 14.2. Suppose you have the four notes, A2, A3, A4 and A5, with corresponding frequencies, 110 Hz, 220 Hz, 440 Hz and 880 Hz.

1. Which note has the longest wavelength? Which note has the shortest wavelength?

   Solution:

2. Compare the number of waves there are in the note A4 as compared to the note A2. [Hint: A4 is 440 Hz and A2 is 110 Hz].

   Solution:

3. How will you write the ratio of the frequencies between A3 and A2? What will this ratio be in lowest terms possible?

   Solution:

4. How will you write the ratio of the frequencies between A4 and A3? What will this ratio be in lowest terms possible?

   Solution:

14.1.1 Activity-to-do: graphing issues—logarithms

Materials needed:

Graph papers of two kinds, linear and semi-log graph paper.

Experiment-to-do:

1. Graph values given in table 14.2 on,
   (a) The linear graph, and
   (b) The semi-log graph. Hint: the semi-log scale, 1, 10, 100, 1000, 10 000…, on the y-axis gives almost a straight line.
2. The values of time should be on the x-axis and values of distance on the y-axis (figures 14.1 and 14.2).

Table 14.2. Activity 1—reference values to plot.

Time (s)	Distance (m)
0	1
1	2
2	4
3	8
4	16
5	32
6	64
7	128
8	256
9	512
10	1024
11	2048
12	4096
13	8192
14	16 384
15	32 768
16	65 536
17	131 072
18	262 144
19	524 288

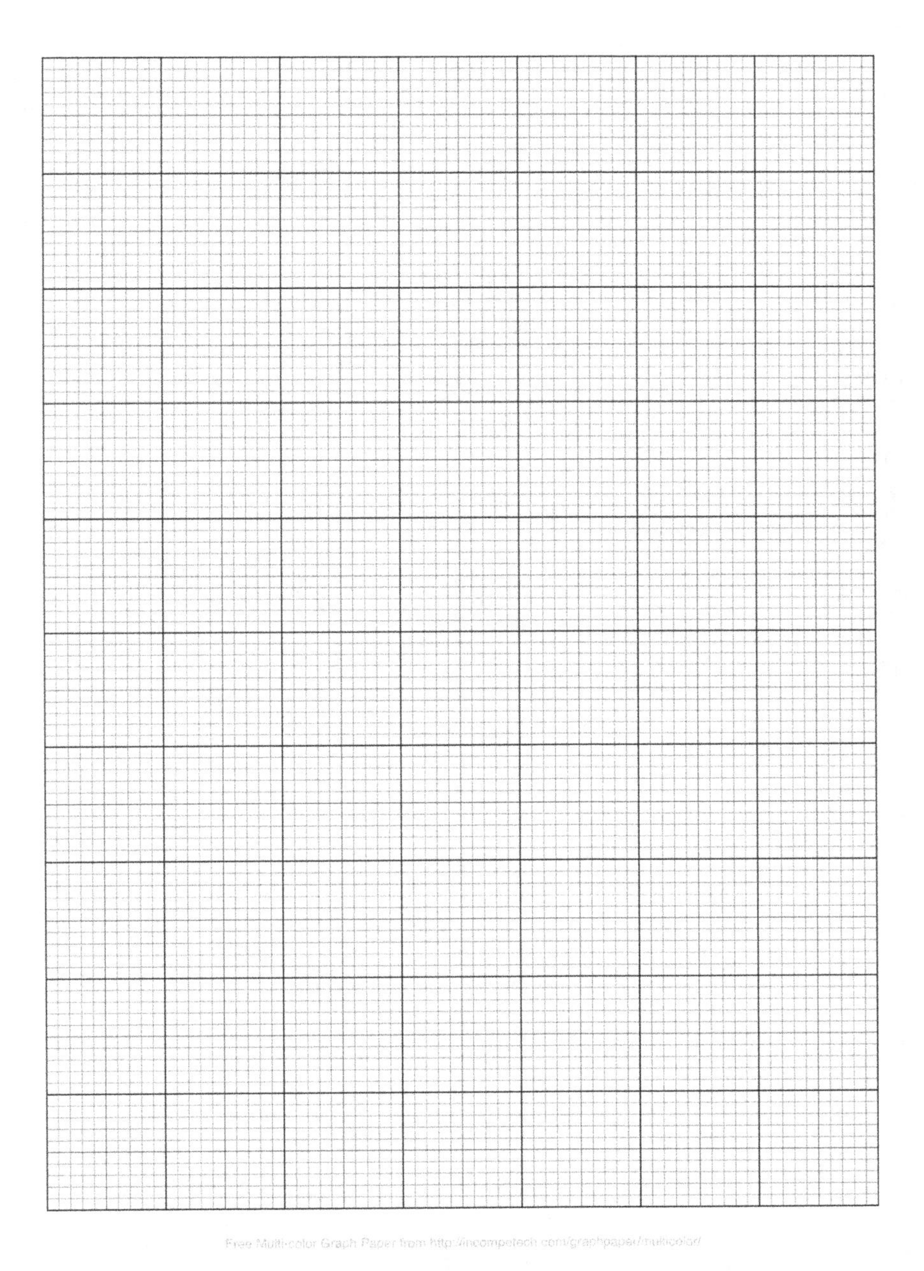

Figure 14.1. Linear graph.

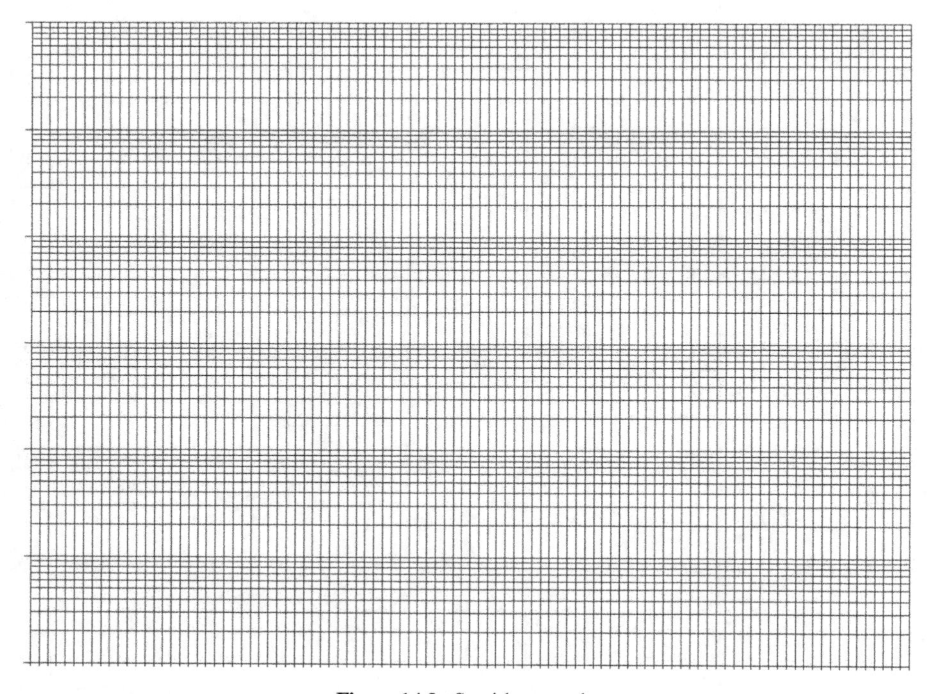

Figure 14.2. Semi-log graph.

14.1.2 Activity-to-do: piano frequencies

Materials needed:

Graph papers; linear and semi-log (figures 14.3 and 14.4).

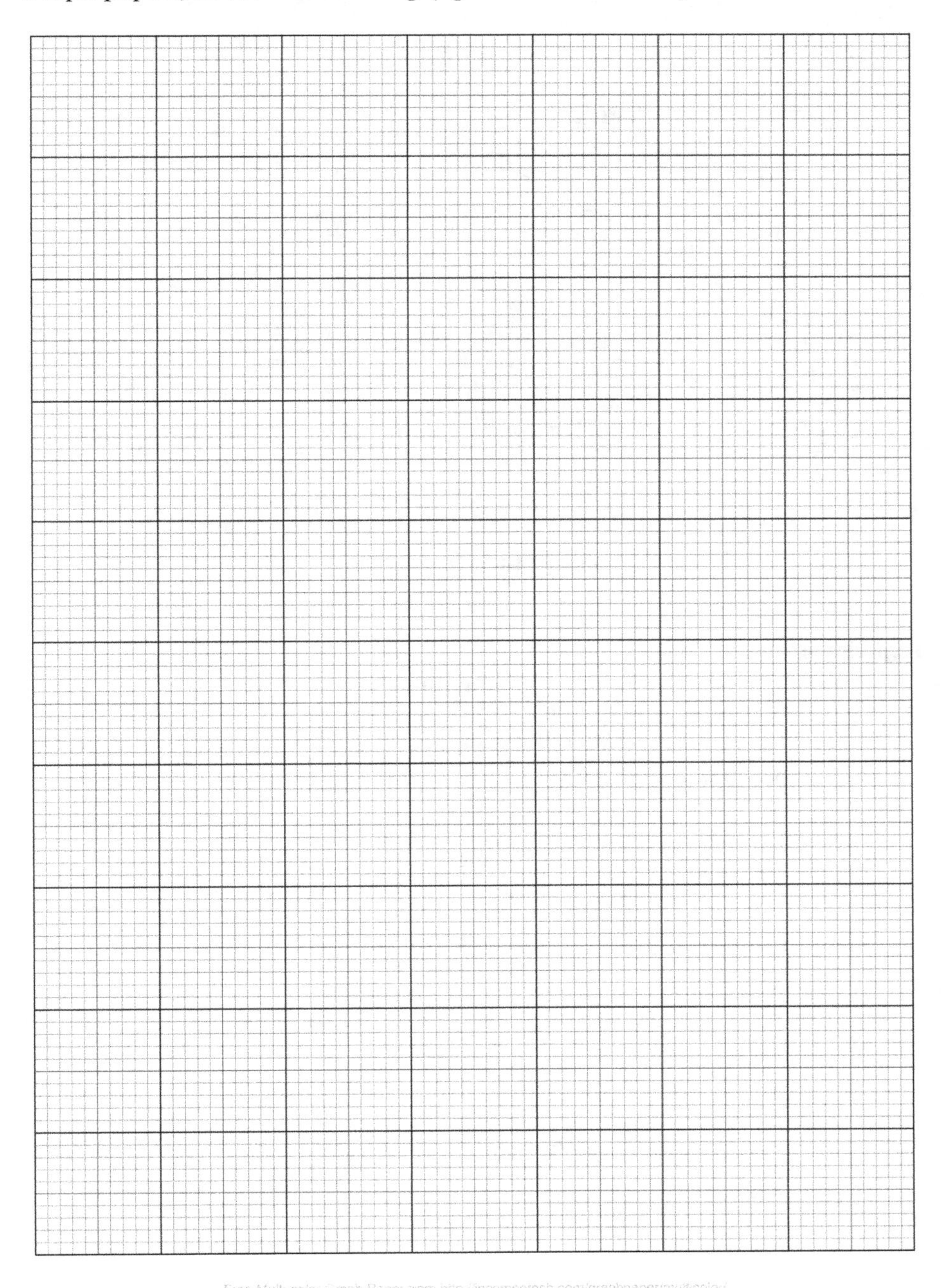

Figure 14.3. Linear graph.

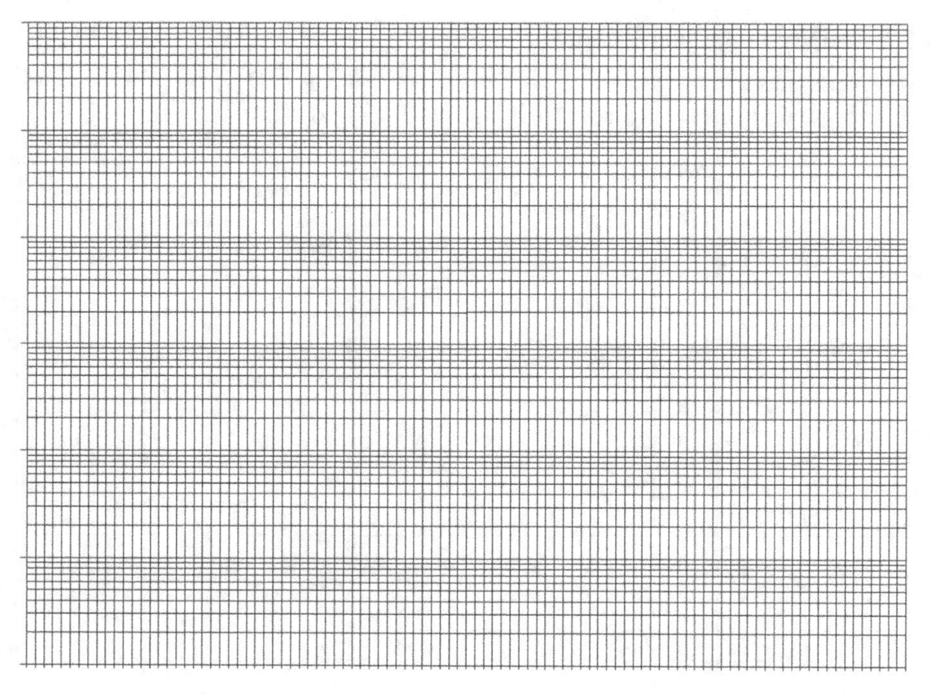

Figure 14.4. Semi-log graph.

Experiment-to-do:

1. Plot values given in tables 14.3 and 14.4.
   (a) the linear graph, and
   (b) the semi-log graph. Hint: the semi-log scale, 1, 10, 100, 1000, 10 000…, on the *y*-axis gives almost a straight line.

Table 14.3. Activity 3—piano frequencies—1.

Piano key	English notation	Frequency (Hz)	Piano key	English notation	Frequency (Hz)
1	A0	27.50	23	G2	97.99
2	A0/B0	29.13	24	G2/A2	103.82
3	B0	30.87	25	A2	110.00
4	C1	32.70	26	A2/B2	116.54
5	C1/D1	34.64	27	B2	123.47
6	D1	36.70	28	C3	130.81
7	D1/E1	38.89	29	C3/D3	138.59
8	E1	41.20	30	D3	146.83
9	F1	43.65	31	D3/E3	155.56
10	F1/G1	46.25	32	E3	164.81
11	G1	48.99	33	F3	174.61
12	G1/A1	51.91	34	F3/G3	184.99
13	A1	55.00	35	G3	195.99
14	A1/B1	58.27	36	G3/A3	207.65
15	B1	61.73	37	A3	220.00
16	C2 (low C)	65.40	38	A3/B3	233.08
17	C2/D2	69.29	39	B3	246.94
18	D2	73.41	40	C4 (middle C)	261.63
19	D2/E2	77.78	41	C4/D4	277.18
20	E2	82.40	42	D4	293.66
21	F2	87.31	43	D4/E4	311.13
22	F2/G2	92.49	44	E4	329.62

Table 14.4. Activity 3—Piano frequencies—2 (continued).

Piano key	English notation	Frequency (Hz)	Piano key	English notation	Frequency (Hz)
45	F4	349.23	67	D6/E6	1244.51
46	F4/G4	369.99	68	E6	1318.51
47	G4	391.99	69	F6	1396.91
48	G4/A4	415.31	70	F6/G6	1479.98
49	A4 (concert pitch)	440.0	71	G6	1567.98
50	A4/B4	466.16	72	G6/A6	1661.22
51	B4	493.88	73	A6	1760.00
52	C5	523.25	74	A6/B6	1864.66
53	C5/D5	554.37	75	B6	1975.53
54	D5	554.37	76	C7	2093.00
55	D5/E5	622.25	77	C7/D7	2217.46
56	E5	659.26	78	D7	2349.32
57	F5	698.46	79	D7/E7	2489.02
58	F5/G5	739.99	80	E7	2637.02
59	G5	783.99	81	F7	2793.83
60	G5/A5	830.61	82	F7/G7	2959.96
61	A5	880.00	83	G7	3135.96
62	A5/B5	932.33	84	G7/A7	3322.44
63	B5	987.76	85	A7	3520.00
64	C6 (high C)	1046.50	86	A7/B7	3729.31
65	C6/D6	1108.73	87	B7	3951.07
66	D6	1174.66	88	C8 (last tone)	4186.01

Part VI

Musical instruments

IOP Publishing

The Physics of Sound and Music, Volume 2
A complete course text (Lab manual)
Samya Bano Zain

Chapter 15

String instruments

15.1 Terms to know about string instruments

1. A string instrument
2. Introduction of energy to an instrument
3. Vibrations in a struck string
4. Changing the pitch in string instruments; by changing material, changing tension in string and length of string
5. The guitar; guitar strings
6. Vibrations of the top plate, back plate and air cavity
7. Electric guitar
8. The piano
9. Bowed string instruments; example, violin
10. Slip-and-stick action.

doi:10.1088/978-0-7503-6350-1ch15

15.2 String instruments

15.2.1 Activity-to-do: pitch experiment—evaluating the frequencies of piano keys

Materials needed:

Oscilloscope; tuning fork; microphone; connector to connect microphone to oscilloscope.

Background information:

An oscilloscope is an electronic measuring instrument that creates a visible two-dimensional graph of one or more electrical potential differences. The horizontal axis of the display represents time and the vertical axis usually shows voltage. The display is caused by a 'spot' that periodically 'sweeps' the screen from left to right. The oscilloscope repeatedly draws a horizontal line called the trace across the middle of the screen from left to right. One of the controls, the timebase control, sets the speed at which the line is drawn, and is calibrated in seconds per division. The vertical control, sets the scale of the vertical deflection, and is calibrated in volts per division.

For a periodic input signal, nearly stable trace is obtained by setting the timebase to match the frequency of the input signal. For example, an input of 50 Hz sine wave has a period (1/50 =) 20 ms, so the timebase should be adjusted so that the time between successive horizontal sweeps is 20 ms. This mode is called continual sweep.

Experiment-to-do:

1. Attach the microphone to the oscilloscope as shown in figure 15.1.
2. Set the oscilloscope as shown on the right of figure 15.1 and complete table 15.1 using different notes on the piano.
    (a) Time: read off the oscilloscope as shown on the right of figure 15.1.
    (b) Remember to convert from 'milli-s' to 's'.

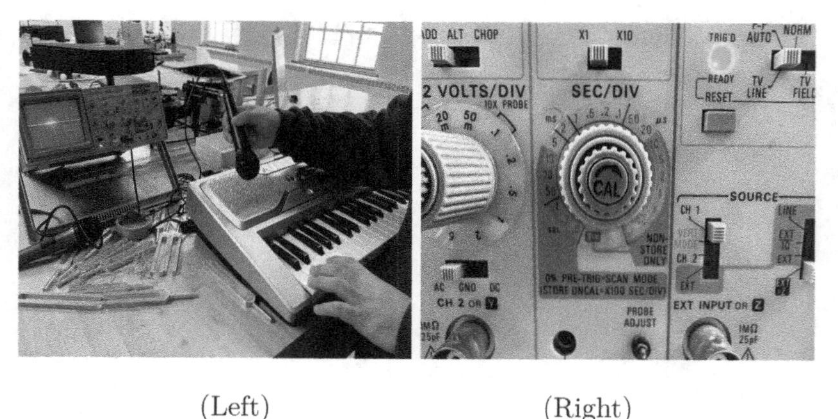

(Left) (Right)

Figure 15.1. (Left): oscilloscope setup. (Right): how to read the time on the oscilloscope for table 15.1, in this case the time is 2 ms.

Table 15.1. Pitch experiment—piano keys.

	Oscilloscope readings						
Piano key	Time (ms)	Time (s)	Distance on x-axis	Calculated period (s)	Evaluated frequency (1/s = Hz).	Actual frequency (Hz)	% difference
Ex 1 (D5)	2	0.002	0.8	= 0.002 × 0.8 = 0.0016	625 Hz	622.25	0.44
Ex 2 (F4)	2	0.002	1.4	= 0.002 × 1.4 = 0.0028	357.14 Hz	349.23	2.26
1							
2							
3							
4							
5							
6							

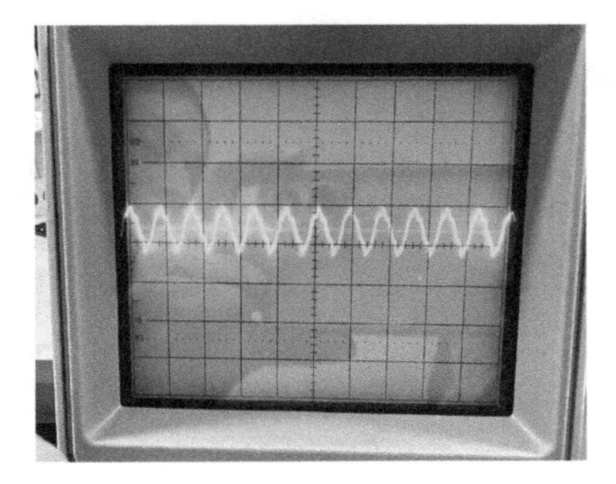

Figure 15.2. To read the distance on x-axis by the oscilloscope, use the distance between two consecutive crests in a wave produced by each note. In this picture the number of boxes are = 0.8, measured from one crest to the next crest.

(c) Distance on x-axis: read the wavelength from crest to crest of each frequency (note) in terms of number of boxes (figure 15.2).
 1 box = 1.0 and each box has 5 divisions.

(d) Calculated period = (time) × (distance on x-axis)

(e) Evaluated frequency,

$$f = \frac{1}{T_P}$$

(f) Some examples of frequencies corresponding to selected note are given in figure 15.3, for more details please refer to tables provided in chapter 14, tables 14.4 and 14.3.

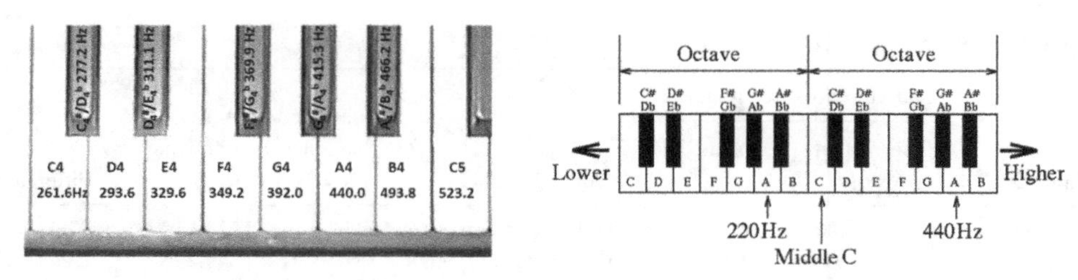

Figure 15.3. Some examples of frequencies that correspond to each note.

Discussion Questions:

1. What was the calculated average % difference?

2. What could be your sources of errors?

3. How could you improve your results?

15.3 Activity-to-do: make your own guitar

Materials needed: Tissue box or a cardboard box; paper towel tube; rubber bands (different sizes and thickness for different sounds); popsicle sticks.

Experiment-to-do:

1. Use a tissue box as the body of the guitar. If you are so inclined, please paint the outside of the box with your favorite color.
2. Also paint the cardboard paper towel tube as you want.
3. Once the paint has dried, attach the painted cardboard paper towel tube to one side of the box and use it as the neck of the guitar. Hint: if you are right handed the neck should be attached to the right side of the guitar when looking down at it and vice versa if you are left handed the neck should be attached to the left side of the guitar.
4. Wrap the rubber bands around the guitar body.
5. Glue three popsicle sticks together to make them thicker, you should have two sets of popsicle sticks.
6. Slide them under the rubber bands and push each set of popsicle stick to either side of the guitar body (outside the opening in the middle of the box).
7. Your guitar is ready to be played! Have fun.

15.4 Soundboards

15.4.1 Activity-to-do: which material makes the best soundboard?

Materials needed:

Sound level meter; a music maker (you may use your phone); same-size samples of various materials to use as soundboards; pieces of foam packing material to be used as supports, a quiet room.

Background information:

All string instruments use a soundboard to amplify the volume of the sound produced by vibrating strings. Soundboards in each instrument are strategically placed such that it gathers the most sound from the vibrating string, amplifies it and then retransmits the amplified sound.

Experiment-to-do:

1. Measure the mass and volume of each soundboard material.
2. Calculate the density (= mass/volume) for each soundboard material.
3. Move to a quiet room.
4. What is the sound level of the music maker alone, without any soundboard at 30 cm? This is your reference sound level.
5. Place each soundboard material on foam packing material, one piece in each of the corners to isolate the soundboard from the table.
6. Place your music maker on the soundboard.
7. Hold the sound level meter at 30 cm from the soundboard and record three readings.
8. Calculate the average dB reading.
9. Subtract this value from the average sound level measured for each soundboard. The result provides you the amount by which the soundboard increased the sound level.
10. Complete table 15.2.
11. Repeat the experiment for each of the soundboard materials.

Table 15.2. Soundboard.

	Reference sound level (dB)	Soundboard readings			Average dB reading (dB)	Amplification by soundboard (dB)
		Trial 1 (dB)	Trial 2 (dB)	Trial 3 (dB)		
Material 1						
Material 2						
Material 3						

Discussion questions:

1. Do you see any relationship between the increased sound level and the material density?

2. Can you predict how the measured sound level may change if the area of the soundboard increases?

3. Alternatively, can you say how the measured sound level will change as the area of the soundboard is decreased?

4. What other properties can you think of that might influence a material's effectiveness as a soundboard?

15.4.2 Activity-to-do: can the players' position affect the ability of a soundboard to transmit sound?

Materials needed:

Any instrument with a soundboard (guitar, violin etc); oscilloscope; sound-level meter; a quiet room.

Experiment-to-do:

In this activity we will see how a players' position affects the ability of a soundboard to transmit sound.

1. Attach the microphone to the oscilloscope.
2. Sit in a chair with your guitar.
3. Place the sound-level meter and the microphone in front of the chair and make sure both point towards the guitar.
4. Please keep the distance between the sound level meter, microphone and guitar constant throughout this experimentation.
5. **Step 1:**
   (a) Hold the guitar snugly so that the back of the guitar body makes contact with both your chest and stomach and strum a chord.
   (b) Pay attention to how hard you are strumming, and try to strum with the same force each time.
   (c) Note the dB reading from the sound level meter. dB reading = __________
   (d) Observe the amplitude of the signal on the oscilloscope. What do you see? Draw or take a picture if it helps.

6. **Step 2:**
   (a) Next, hold the guitar such that the bottom of guitar rests on your right thigh and that your right forearm and upper chest make contact with the top part of the guitar's body. There should be about 6" of space between the soundboard and your torso.
   (b) Strum the same chord as step 1 with approximately the same force.
   (c) Note the dB reading from the sound level meter. dB reading = __________
   (d) Observe the amplitude of the signal on the oscilloscope. What do you see? Draw or take a picture if it helps.

7. **Step 3:**
   (a) For step 3, your left foot should be elevated, say by resting it on a foot stool.
   (b) Hold the guitar such that the bottom of the guitar rests on your left thigh with your right forearm resting on the top of the guitar body. *Tilt only the body of the guitar back* until the top of the guitar touches your chest. Please ensure that there are about 6” of space between the soundboard and your chest.
   (c) Strum the same chord (with the same force) as step 1.
   (d) Note the dB reading from the sound level meter. dB reading = __________
   (e) Observe the amplitude of the signal on the oscilloscope. What do you see? Draw or take a picture if it helps.

8. **Step 4:**
   (a) For step 4, hold the guitar such that the bottom of the guitar rests on your right thigh and your right forearm rests on the top of the guitar body. This time make sure that the entire back of the guitar touches your chest.
   (b) Strum the same chord (with the same force) as step 1.
   (c) Note the dB reading from the sound level meter. dB reading = __________
   (d) Observe the amplitude of the signal on the oscilloscope. What do you see? Draw or take a picture if it helps.

Discussion questions:

1. In which position do you have the least contact with the guitar’s soundboard?

2. Which position produced the greatest amplitude of sound?

3. Alternatively, which position produced the greatest damping of sound?

15.5 Activity-to-do: singing wineglasses

Materials needed:

Wine glasses; water.

Background information:

When you tap the rim of a glass, you hear a sound because tapping sets the glass into vibration, generally at its resonance frequency. The pitch of sound depends on a number of factors, like the size and shape of the glass. The rim and sides of the glass vibrate together. The sides of the glass also vibrate and the shape of the glass changes, often several hundred times per second. Slip–stick motion occurs when the right amount of frictional force causes one object to alternate between slipping over another object and sticking to it.

Slip–stick motion occurs when the right amount of frictional force causes one object to alternate between slipping over another object and sticking to it. The finger motion can start the vibration in glass and the sliding motion rapidly pushes and releases the glass continuously which transfers energy from the motion of your finger into the glass. The slip–stick motion appears to be uniform, smooth and continuous, even though it really is not.

Experiment-to-do:

1. Set the wine glasses in a line.
2. Keep one glass empty and in the others add water to different levels.
3. Rub your finger around the rim of the empty glass to produce a sound. When you rub your finger along the rim of a glass, you produce the vibration pattern at the natural frequency of the glass. Hint: wetting your finger makes the slip–stick motion easier to produce.
4. Explain in your own words what is happening in terms of '*slip–stick*' motion and friction.

5. Put a ping-pong ball in the bottom of the empty glass and rub your finger around the rim to make the glass sing. Do you notice any movement with the ping-pong ball?

6. You can make the sound louder or softer, by changing the speed and/or applied pressure? Does doing this change the vibration frequency?

7. One way to see the vibration pattern is to fill a glass almost full of water and rub your finger around the rim. Are the ripples produced larger at some points than at others? Explain in your own words what you observe.

8. Add water to the other glasses to adjust the pitch. Rub your finger around the rims. Does the water in the glass stay still or do you notice any movement?

9. Does the pitch change with different water levels?

10. Can you find a relationship between the water level in the glass and the resulting pitch of the note?

11. By adding different amounts of water to different glasses, you can produce a whole series of different musical notes and play songs and hence will make your own ‘glass harmonica’.

12. Repeat the experiment with different sizes or shapes of glasses. Does the size or shape of the glass change the sounds or pitches produced?

IOP Publishing

The Physics of Sound and Music, Volume 2

A complete course text (Lab manual)

Samya Bano Zain

Chapter 16

Percussion instruments

16.1 Terms to know about percussion instruments

A percussion instrument is any object which produces a sound when it is hit, or shaken, or rubbed, or scraped, or by any other way which sets the object into vibration. Percussion instruments include tuning forks, chimes, xylophones, marimbas etc. The earliest known drum, made from an elephant skin was discovered preserved in Antarctica's ice age.

1. Vibrations in a bar
2. Vibrations in plates and membranes
3. Chladni plate
4. Rhythmic percussion
5. Membranophone
6. Idiophones
7. Electrophones
8. Vibrations of bells.

doi:10.1088/978-0-7503-6350-1ch16

16.2 Activity-to-do: class activity: homemade xylophone (pipe)

Materials needed:

Tone-O-phone; rubber bands; mallet; homemade xylophone kit or electrical metal tubing, cut into lengths (28.6 cm, 28 cm, 26 cm, 25.72 cm, 24 cm, 23 cm, 22.2 cm, 21 cm, 20 cm, 18.4 cm, 18 cm, 17 cm, 16 cm) (figure 16.1).

Experiment-to-do:

1. Place the pipes from longest to shortest on the table.
2. Hit with the mallet to hear the sound each pipe makes.
3. What do you observe?

4. Adjust the pitch of the pipes, as needed, by moving the rubber bands.
5. Play a song on the provided tone-O-phone.
6. Make your own xylophone by using the metal tubing.
7. What did you have to do to accomplish this task?

8. Please play the same song on both the xylophone you made and the tone-O-phone to sound the same. Were you successful?

9. What is the phenomenon behind moving the rubber bands and the change in pitch for the pipe xylophone?

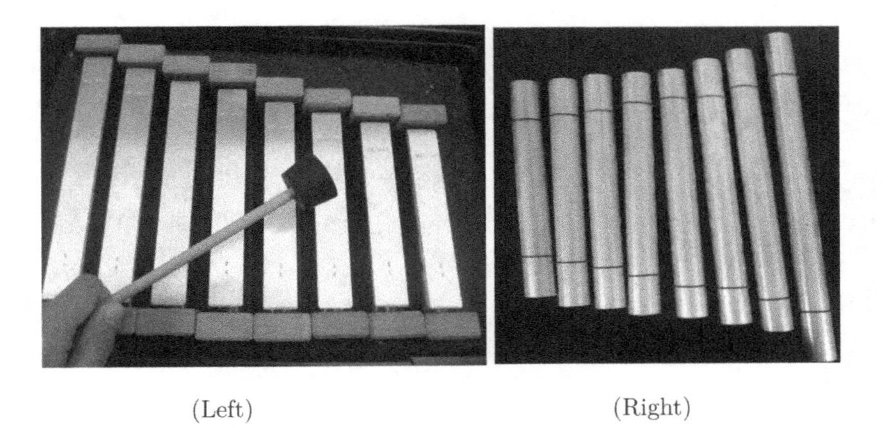

(Left) (Right)

Figure 16.1. (Left) tone-O-phone. (Right) homemade xylophone using metal tubing.

16.3 Activity-to-do: class activity: bottle xylophone

Materials needed:

Glass bottles (Snapple bottles work well), emptied and clean; metal spoon or thick wooden dowel (figure 16.2).

Experiment-to-do:

1. Fill each bottle with different amounts of water.
2. Tap on the bottles using a spoon to find the pitches.
3. See if you can repeat the same song as in section 16.2.
4. Draw what happens to the sound wave as you increase the amount of water in the bottle.

Figure 16.2. Homemade bottle xylophone using glass bottles and water.

Discussion questions:

1. What is the phenomenon behind changing the water level and the change in pitch for the water xylophone?

2. What happens to the wavelength (λ) as water level rises?

3. What happens to frequency (f) as water level rises? (keep in mind that $v = \lambda f$ where v, the speed of sound is a constant)

4. What happens to pitch as water level rises?

16.4 Activity-to-do: class activity: make your own kazoo

Materials needed:

Empty paper towel tube; plastic sheet (5" × 5" square); aluminum foil (5" × 5" square); parchment paper or wax paper (5" × 5" square); rubber band; scissors; pencil.

Background information:

A kazoo is a very simple musical instrument, made up of a hollow pipe with a hole and one end covered by a membrane to form a closed end. Even though a kazoo looks like a flute it is actually most like a drum, hence it is placed in the chapter about percussion instruments. As the player sings or hums into the open end of the kazoo, sound waves travel through the kazoo and strike the membrane which causes it to vibrate adding resonance to the original sound. Additionally, as the sound travels through the tube, some waves bounce off the walls and add harmonics to the sound. Both the resonance and harmonics create the characteristic buzzing of the kazoo. Remember, resonances are identical to the initial sound wave and hence they increase its volume, whereas harmonics are not identical to the initial sound wave and hence they change the timbre of the sound (figure 16.3).

Experiment-to-do:

1. **Step 1:**
   (a) Say 'Kaa-Zooo'. Listen to your voice.
   (b) How does your voice sound?

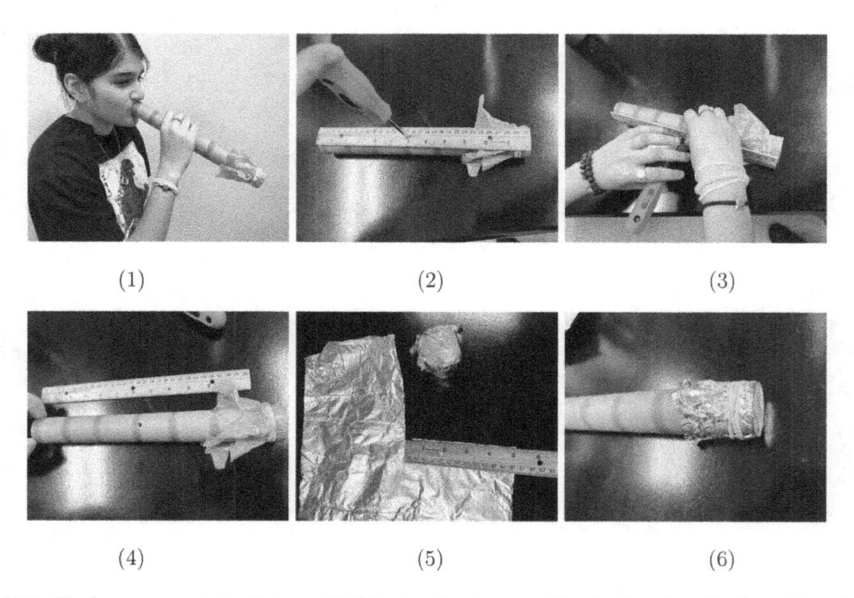

Figure 16.3. To kazoo or not to Kazoo. (1) Playing the kazoo with plastic edge, (2–3) making the hole, (4) cutting the aluminum foil, (6) kazoo with the foil.

2. **Step 2:**
   (a) Put one end of the empty paper towel tube to your mouth such that it touches the skin above and below your mouth.
   (b) Say 'Kaa-Zooo' into the empty paper towel tube. Listen to your voice.
   (c) Does your voice sound different as it travels through the tube? What is different about it?

   (d) Can you feel the tube vibrating as you speak?

3. **Step 3:**
   (a) Cut a (4" × 4" square) of plastic sheet. Secure it with a rubber band on the other end of the empty paper towel tube.
   (b) Say 'Kaa-Zooo' again. Listen to your voice.
   (c) Does your voice sound different as it travels through the tube? What is different about it?

4. **Step 4:**
   (a) Next, use a sharpened pencil or some other sharp tool (screwdriver etc) to poke a hole halfway between the two ends of the empty paper towel tube.
   (b) Say 'Kaa-Zooo' again. Listen to your voice.
   (c) Does your voice sound different as it travels through the tube? What is different about it?

   (d) What happens when you cover and uncover the hole with your finger as you speak?

5. Repeat step 4 using aluminum foil to cover the end of the tube. How does the sound of your voice change with the aluminum foil covering?

6. Repeat step 4 using a parchment paper or wax paper to cover the end of the tube. How does the sound of your voice change with the parchment paper or wax paper covering?

IOP Publishing

The Physics of Sound and Music, Volume 2
A complete course text (Lab manual)
Samya Bano Zain

Chapter 17

Wind instruments

17.1 What to know about wind instruments

1. Wind instruments
2. Sound production in humans
3. Standing waves in pipes
4. Pipe–reed system
5. Instruments in the brass family
6. Production of sound in brass instruments
7. Timbres of different instruments.

doi:10.1088/978-0-7503-6350-1ch17

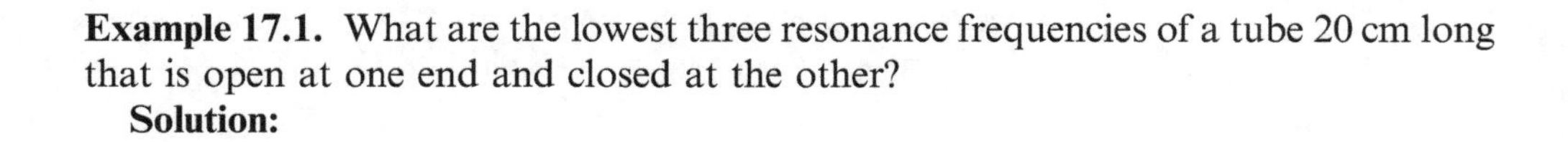

Example 17.1. What are the lowest three resonance frequencies of a tube 20 cm long that is open at one end and closed at the other?

Solution:

Example 17.2. What are the lowest three resonance frequencies of a tube 20 cm long that is open at both ends?

Solution:

17.1.1 Activity-to-do: class activity: pan flute: what factors affect the pitch of wind instruments?

Materials needed:

6–7 plastic drinking straws per student; ruler; scissors; clear tape.

Experiment-to-do:

Make a straw pan flute using a few drinking straws and tape.

1. Cut the plastic drinking straws; you should have at least 6 straws in the following lengths: 2”, 3”, 4”, 5”, 6” and 7”.
2. Place the straws from longest to shortest on the table.
3. Cut across one end of each straw as shown in the figure 17.1.
4. Once the straws are cut, pick each one and blow across the uncut end by keeping your lips and chin close to the straw as shown in figure 17.2.

(Left) (Right)

Figure 17.1. Making a pan flute.

Figure 17.2. Playing the flute.

5. What do you observe as you blow across each from longest to shortest?

6. Place all the straws on the table and tape them from the largest to smallest. You have just made your own pan flute!
7. Try to play a song on your pan flute.
8. **Extra challenge:** see if you are able to play the same song on your hand-made pan flute, your hand-made xylophone and the tone-O-phone! Were you successful?

Discussion questions:

1. How does the length of the air column or windpipe determine the pitch?

2. What is the phenomenon behind the observed change in pitch?

17.1.2 Activity-to-do: class activity: make a plastic straw flute

Materials needed:

Two plastic drinking straws of different lengths (hint: thickness of the straw has an effect, so get the really thin straws); scissors.

Experiment-to-do:

1. Hold the plastic straw with the smaller length and press one end about 2 cm until it is creased. You may use your teeth or pinch it between your fingers or fingernails to flatten it.
2. Once the end is creased, use the scissor to cut the end twice so it results in a point and then cut the point off so it forms a flat edge, as shown in the figure 17.3.
3. This will form the mouthpiece of your flute. The two cut edges of the straw act like a reed.
4. Put the pointy cut end of the straw just behind your lips (please be careful!). Then curve your lips down and inward a little and apply light pressure on the straw with your lips and blow gently and softly and increase the pressure until the straw makes a sound. The cut ends should vibrate and produce a tone. If you do not hear the sound slowly move the point back and forth, and/or increase or decrease the pressure on the straw until you do. Usually, most people blow too hard at first so you may have to be very gentle at the first go.
5. Similarly cut the other straw so both straws have angular cuts at one end.

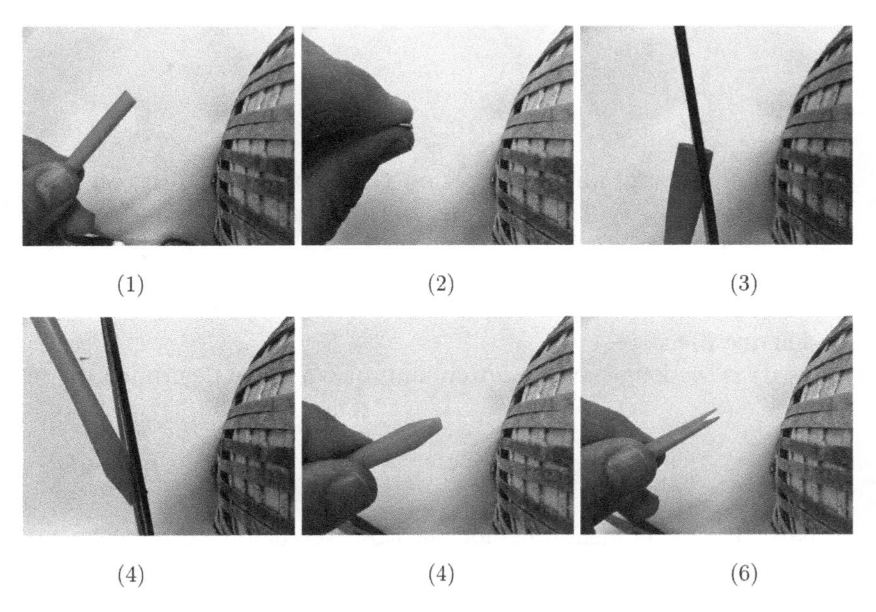

Figure 17.3. Make a straw flute. From left. (1) Take a plastic straw. (2) Pinch one end. (3) Cut one side. (4) Cut the other side. (5) Final flute (top view) (6) Final flute (side view).

6. Blow into the shorter straw. How does it sound? Note the pitch.

7. Try the same experiment with the longer straw using the same method outlined in step 4. Again, you may need to try blowing through the straw a few times to make it produce a constant, single note. What do you observe? Is there a change in pitch?

8. **Variation in the experiment**
   (a) Change the physical length of the straw, by cutting one end by one inch. What do you observe?

   (b) Change how hard you blow through the straws. What do you observe?

   (c) Try different angles of cuts on your straw. What do you observe?

Discussion questions:

1. How does the length of the air column or windpipe determine the pitch?

2. What is the phenomenon behind the observed change in pitch?

17.1.3 Activity-to-do: class activity: make a straw slide trombone

Materials needed:

Two plastic straws of different diameters (hint: thickness of the straw has an effect, so get the really thin straws); scissors.

Experiment-to-do:

Try the same experiment as performed in activity 17.1.2 with larger bendy straws to create a slide trombone by joining the two straw types together!

1. Hold the straw with the smaller diameter and press one end until it is creased a bit.
2. Once the end is creased, use the scissor to cut the end twice so it results in a point and then cut the point off so it forms a flat as shown in figure 17.3.
3. Slide the smaller straw into the larger straw. The two straws must be able to slide somewhat tightly around each other but not be too loose.
4. Put the pointy cut end of the smaller diameter straw just behind your lips (please be careful!). Then curve your lips down and inward a little and apply light pressure on the straw with your lips and blow gently and softly and increase the pressure until the straw makes a sound. If you do not hear the sound slowly move the point back and forth, and/or increase or decrease the pressure on the straw until you do. Usually, most people blow too hard at first so you may have to be very gentle at the first go.
5. Once you do hear the sound, slide the large straw back and forth. What do you observe as you change the length by moving the straws over each other?

Printed in the USA
CPSIA information can be obtained
at www.ICGtesting.com
CBHW081240060824
12632CB00004B/14

9 780750 363488